This edition of Joint Publication (JP) 5-0, *Joint Planning*, reflects current doctrine for conducting joint, interagency, and multinational planning activities across the range of military operations. This keystone publication is part of the core of joint doctrine and establishes the planning framework for our forces' ability to fight and win as a joint team.

As our military continues to serve and protect our Nation in the complex environment of global competition and conflict, we must continually refine our doctrine and update our planning practices based upon those experiences and lessons learned. Our understanding of operations across the spectrum of conflict and the information needed by senior leaders to make strategic and operational-level decisions, developed during the planning process has evolved. This update to JP 5-0 ensures all our operations benefit from the application of our doctrinal planning processes.

Likewise, the practice of Adaptive Planning and Execution has continued to evolve since the last publication of JP 5-0. This publication provides necessary updates to that process, as our combatant commands have continued to develop the ability to provide military options for contingencies. Therefore, we seek to develop tools that allow for more rapid development, review, and refinement of plans at the accelerated pace the world requires today.

Given that the operational environment is not simple or static, adaptation and flexibility are necessary in planning and execution. This edition of JP 5-0 seeks to provide joint force commanders and their component commanders with processes that allow for that flexibility and the ability to plan and develop plans for an uncertain and challenging environment.

Our Armed Forces serve to support our national leadership in attaining national objectives. I encourage leaders to ensure their organizations understand and use joint doctrine and this Joint Publication in particular as you continue to assist our Nation in advancing its enduring interests.

For the Chairman of the Joint Chiefs of Staff:

KEVIN D. SCOTT
Vice Admiral, USN
Director, Joint Force Development

PREFACE

1. Scope

This publication is the keystone document for joint planning. It provides the doctrinal foundation and fundamental principles that guide the Armed Forces of the United States in planning joint campaigns and operations.

2. Purpose

This publication has been prepared under the direction of the Chairman of the Joint Chiefs of Staff (CJCS). It sets forth joint doctrine to govern the activities and performance of the Armed Forces of the United States in joint operations, and it provides considerations for military interaction with governmental and nongovernmental agencies, multinational forces, and other interorganizational partners. It provides military guidance for the exercise of authority by combatant commanders and other joint force commanders (JFCs), and prescribes joint doctrine for operations and training. It provides military guidance for use by the Armed Forces in preparing and executing their plans and orders. It is not the intent of this publication to restrict the authority of the JFC from organizing the force and executing the mission in a manner the JFC deems most appropriate to ensure unity of effort in the accomplishment of objectives.

3. Application

a. Joint doctrine established in this publication applies to the Joint Staff, commanders of combatant commands, subordinate unified commands, joint task forces, subordinate components of these commands, the Services, and combat support agencies.

b. The guidance in this publication is authoritative; as such, this doctrine will be followed except when, in the judgment of the commander, exceptional circumstances dictate otherwise. If conflicts arise between the contents of this publication and the contents of Service publications, this publication will take precedence unless the CJCS, normally in coordination with the other members of the Joint Chiefs of Staff, has provided more current and specific guidance. Commanders of forces operating as part of a multinational (alliance or coalition) military command should follow multinational doctrine and procedures ratified by the United States. For doctrine and procedures not ratified by the US, commanders should evaluate and follow the multinational command's doctrine and procedures, where applicable and consistent with US law, regulations, and doctrine.

Intentionally Blank

- **Changes title from "Joint Operation Planning" to "Joint Planning."**

- **Adds a chapter to introduce a campaign planning concept to organize and direct daily operations outside of combat.**

- **Identifies the requirement to provide multiple feasible options at the combatant command level to fulfill Chairman of the Joint Chiefs of Staff and Secretary of Defense decision processes.**

- **Adds the requirement to identify operational and strategic risk for decision makers.**

- **Links campaign planning and execution to contingency planning.**

  - **Identifies contingency plans as branches and sequels to campaign plans.**

  - **Identifies the assessments of campaign plans impact on assumptions and conditions for contingency plans.**

- **Removes "deliberate" and "crisis action" planning terms as both use the same processes.**

- **Removes the six-phase phasing model, but does not change the definition of phases or the use of phasing as a planning tool.**

- **Updates and expands the discussion of assessments.**

- **Expands the discussion on risk.**

- **Adds appendices on posture plans, theater distribution plans, and red teams.**

- **Updates terms and definitions.**

Intentionally Blank

TABLE OF CONTENTS

CHAPTER III
STRATEGY AND CAMPAIGN DEVELOPMENT

CHAPTER IV
OPERATIONAL ART AND OPERATIONAL DESIGN

Section A: Operational Art

Section B: Operational Design

Section C. Elements of Operational Design

Section D. Phasing

GLOSSARY

FIGURE

- **Describes the principles of joint planning.**

- **Discusses Strategy, Strategic Art, Operational Art, and Operational Planning.**

- **Compares Strategic, Theater, and Functional Planning.**

- **Discusses Strategic Guidance and Coordination and the Joint Strategic Planning System.**

- **Outlines Strategy and Campaign Development.**

- **Describes the Joint Planning Process.**

- **Discusses Operation Assessment.**

- **Explains conditions for transitioning planning to execution.**

Joint Planning

Joint planning is the deliberate process of determining how (the **ways**) to use military capabilities (the **means**) in time and space to achieve objectives (the **ends**) while considering the associated **risks.** Ideally, planning begins with specified national strategic objectives and military end states to provide a unifying purpose around which actions and resources are focused.

At the combatant command (CCMD) level, joint planning serves two critical purposes:
- At the strategic level, joint planning provides the President and the Secretary of Defense (SecDef) options, based on best military advice, on use of the military in addressing national interests and achieving the objectives in the *National Security Strategy* (NSS) and Defense Strategy Review (DSR).
- At the operational level, once strategic guidance is given, planning translates this guidance into specific activities aimed at achieving strategic and operational-level objectives and attaining the military end state.

Principles of Planning

Focuses on the End State. Joint planning is end state oriented: plans and actions positively contribute to achieving national objectives.

Globally Integrated and Coordinated. Planning considers that operations take place throughout the operational environment (OE) irrespective of geographic, political, or domain boundaries.

Resource Informed. Joint planning provides a realistic assessment of the application of forces, given current readiness, availability, location, available transportation, and speed of movement.

Risk Informed. Planning provides decision makers an honest assessment of the costs and potential consequences of military actions.

Framed within the OE. Adaptive planning is based on continuous monitoring and understanding of actual conditions affecting the OE such as current friendly and adversary force postures, readiness, geopolitical conditions, and adversary perceptions.

Informs Decision Making. Planning identifies issues and assumptions required for planning to continue, likely resource requirements, costs and cost-benefit trade-offs, and risks associated with different courses of action (COAs).

Adaptive and Flexible. Planning occurs in a networked, collaborative environment that requires dialogue among senior leaders; concurrent plan development; and collaboration across strategic, operational, and tactical planning levels.

Strategy, Strategic Art,
Operational Art, and
Operational Planning

Strategy is a prudent idea or set of ideas for employing the instruments of national power in a synchronized and integrated fashion to achieve theater, national, and/or multinational objectives. Strategy can also be described as the art and science of determining a future state/condition (ends), conveying this to an audience, determining the operational approach (ways), and identifying the authorities and resources (time, forces, equipment, money, etc.) (means) necessary to reach the intended end state, all while managing the associated risk.

Strategy should not be confused with strategic-level guidance.

Strategic art is the ability to understand the strategic variable and to conceptualize how the desired objectives set forth in strategic-level guidance can be reached through the employment of military capabilities.

Operational art is the application of intuition and creative imagination by commanders and staffs.

Operational planning translates the commander's concepts into executable activities, operations, and campaigns, within resource, policy, and national limitations to achieve objectives.

Strategic, Theater, and Functional Planning

Combatant commanders (CCDRs) use strategic guidance and direction to prepare command strategies, focused on their command's specific capabilities and missions to link national strategic guidance to theater or functional strategies and joint operations. The command strategy, like national strategy, identifies broad, long-range objectives the command aims to achieve as their contribution toward national security. The command strategy provides the link between national strategic guidance and joint planning.

Strategy, Plans, Operations, and Assessments Cycle

Plans translate the strategy into operations with the expectation that successful operations achieve the desired strategic objectives. Similarly, the effects of operations, successful or otherwise, change the operational and strategic environment, requiring constant evaluation of the strategic-level objectives to ensure they are still relevant and feasible. Joint forces, through their assessments, identify when their actions begin to negatively affect the OE, and change their operations and activities to ensure better alignment between the actions and objectives.

Shared Understanding

Civilian-Military Dialogue. Strategy and joint planning occur within Adaptive Planning and Execution (APEX), the department-level enterprise of policies, processes, procedures, and reporting structures supported by communications and information technology used by the joint planning and execution community (JPEC) to plan and execute joint operations. A focus of APEX is the interaction between senior Department of Defense (DOD) civilian leadership, CCDRs, and Chairman of the Joint Chiefs of Staff (CJCS), which helps the President and SecDef decide when, where, and how to employ US military forces and resources.

Agility, Initiative, and Simplicity

The key tenets of a plan–the commander's mission, intent, and objectives are likely to endure, subject to changes in policy and/or strategy. Operation assessment provides the means to review their validity, and reaffirm or adjust as necessary. Meanwhile, based on continuous operation assessment, the scheme of maneuver (including supporting effects and planned activities) and main effort are likely to be refreshed more frequently as the plan progresses and the command seeks to maintain the initiative.

Interorganizational Planning and Coordination

Interorganizational planning and coordination is the interaction that occurs among elements of DOD; participating US Government departments and agencies; state, territorial, local, and tribal agencies; foreign military forces and government departments and agencies; international organizations; nongovernmental organizations; and the private sector for the purpose of accomplishing an objective.

Multinational Planning and Coordination

Joint planning will frequently be accomplished within the context of multinational planning. There is no single doctrine for multinational action, and each multinational force develops its own protocols, operation plans (OPLANs), concept plans, and operation orders (OPORDs). US planning for multinational operations should accommodate and complement such protocols and plans.

Strategic Guidance and Coordination

National and Department of Defense (DOD) Guidance

The President, SecDef, and CJCS provide their orders, intent, strategy, direction, and guidance via strategic direction to the military to pursue national interests within legal and constitutional limitations. They generally communicate strategic direction to the military through written documents, but it may be communicated by any means available. Strategic direction is contained in key documents, generally referred to as strategic guidance.

Department of State and the United States Agency for International Development

The Department of State (DOS) is the lead US foreign affairs agency within the Executive Branch and the lead institution for the conduct of American diplomacy. The Secretary of State is the President's principal foreign policy advisor. The Secretary of State implements the President's foreign policies worldwide through DOS and its employees. US Agency for International Development is an independent federal agency that receives overall foreign policy guidance from the Secretary of State.

DOD

DSR. The DSR articulates a defense strategy consistent with the most recent NSS by defining force structure, modernization plans, and a budget plan allowing the military to successfully execute the full range of missions within that strategy for the next 20 years.

Unified Command Plan (UCP). The UCP, signed by the President, establishes CCMD missions and CCDR responsibilities, addresses assignment of forces, delineates geographic area of responsibility for geographic CCDRs, and specifies responsibilities for functional CCDRs.

***Guidance for Employment of the Force* (GEF).** The GEF, signed by SecDef, and its associated Contingency Planning Guidance, signed by the President, convey the President's and SecDef's guidance for contingency force management, security cooperation, and posture planning. The GEF translates NSS objectives into prioritized and comprehensive planning guidance for the employment of DOD forces.

***Global Force Management Implementation Guidance* (GFMIG).** The GFMIG aligns force assignment, apportionment, and allocation methodologies in support of the DSR and GEF, joint force availability requirements, and joint force assessments.

Joint Strategic Planning System

The Joint Strategic Planning System (JSPS) is the primary system by which the CJCS carries out statutory responsibilities. The JSPS enables the CJCS to conduct assessments; provide military advice to the President, SecDef, National Security Council, and Homeland Security Council and assist the President and SecDef in providing strategic direction to the US Armed Forces.

Strategic Direction. The President, SecDef, and CJCS use strategic direction to communicate their broad goals and issue-specific guidance to DOD. It provides the common thread that integrates and synchronizes the planning activities and operations of the Joint Staff, CCMDs, Services, joint forces, combat support agencies (CSAs), and other DOD agencies.

National Military Strategy **(NMS).** The NMS, derived from the NSS and DSR, prioritizes and focuses the efforts of the Armed Forces of the United States while conveying the CJCS's direction with regard to the OE and the necessary military actions to protect national security interests.

Joint Strategic Campaign Plan (JSCP). The JSCP provides military strategic and operational guidance to CCDRs, Service Chiefs, CSAs, and applicable DOD agencies for preparation of plans based on current military capabilities.

Combatant Commanders

At the CCMD level, a joint planning group, operational planning group, or operational planning team is typically established to direct planning efforts across the command, including implementation of plans and orders.

Strategic Estimate. The CCDR and staff, with input from subordinate commands and supporting commands and agencies, prepare a strategic estimate by analyzing and describing the political, military, economic, social, information, and infrastructure factors and trends, and the threats and opportunities that facilitate or hinder achievement of the objectives over the timeframe of the strategy.

CCMD Strategies. A strategy is a broad statement of the commander's long-term vision. It is the bridge between national strategic guidance and the joint planning required to achieve national and command objectives and attain end states.

Commander's Communication Synchronization

Commander's communication synchronization is the process to coordinate and synchronize narratives, themes, messages, images, operations, and actions to ensure their integrity and consistency to the lowest tactical level across all relevant communication activities.

Application of Guidance

The **supported CCDR** has primary responsibility for all aspects of a task assigned by the GEF, the JSCP, or other joint planning directives. In the context of joint planning, the supported commander can initiate planning at any time based on command authority or in response to direction or orders from the President, SecDef, or CJCS.

Supporting commanders provide forces, assistance, or other resources to a supported commander. Supporting commanders prepare supporting plans as required.

Adaptive Planning and Execution Enterprise

APEX integrates the planning activities of the JPEC and facilitates the transition from planning to execution. The APEX enterprise operates in a networked, collaborative environment, which facilitates dialogue among senior leaders, concurrent and parallel plan development, and collaboration across multiple planning levels.

Operational Activities

Operational activities are comprised of a sustained cycle of situational awareness, planning, execution, and assessment activities that occur continuously to support leader decision-making cycles at all levels of command.
- Situational awareness addresses procedures for describing the OE, including threats to national security.
- Planning translates strategic guidance and direction into campaign plans, contingency plans, and OPORDs.
- Execution begins when the President or SecDef authorizes the initiation of a military operation or other activity. An execute order, or other authorizing directive, is issued by the CJCS at the direction of the President or SecDef to initiate or conduct the military operations.
- Assessment determines the progress of the joint force toward mission accomplishment. Throughout the four planning functions, assessment involves comparing desired conditions of the OE with actual conditions to determine the overall effectiveness of the campaign or operation.

Planning Functions

The four planning functions of strategic guidance, concept development, plan development, and plan assessment are generally sequential, although often run simultaneously in order to deepen the dialogue between civilian and military leaders and accelerate the overall planning process. SecDef, CJCS, or the CCDR may direct the planning staff to refine or adapt a plan by reentering the planning process at any of the earlier functions.

Strategic Guidance. Strategic guidance initiates planning, provides the basis for mission analysis, and enables the JPEC to develop a shared understanding of the issues, OE, objectives, and responsibilities.

Concept Development. During planning, the commander develops several COAs, each containing an initial concept of operations (CONOPS) that should identify major capabilities and authorities required and task organization, major operational tasks to be accomplished by components, a concept for employment and sustainment, and assessment of risk.

Plan Development. This function is used to develop a feasible plan or order that is ready to transition into execution.

Plan Assessment (Refine, Adapt, Terminate, Execute). Commanders continually review and evaluate the plan; determine one of four possible outcomes: refine, adapt, terminate, or execute; and then act accordingly.

Planning Products

Campaign Plans. A campaign is a series of related military operations aimed at accomplishing strategic and operational objectives within a given time and space.

Contingency Plans. Contingency plans are branches of campaign plans that are planned for potential threats, catastrophic events, and contingent missions without a crisis at-hand, pursuant to the strategic guidance in the UCP, GEF, published strategic guidance statements, the JSCP, and of the CCDR.

Supporting Plans. Supporting CCDRs, subordinate joint force commanders (JFCs), component commanders, and CSAs prepare supporting plans as tasked by the JSCP or other planning guidance.

Products of Planning in Crises

Planning initiated in response to an emergent event or crisis uses the same construct as all other planning.

However, steps may be compressed to enable the time-sensitive development of OPLANs or OPORDs for the deployment, employment, and sustainment of forces and capabilities in response to a situation that may result in actual military operations.

Strategy and Campaign Development

The CCDR's strategy prioritizes the ends, ways, and means within the limitations established by the budget, global force management processes, and strategic guidance/direction. Strategy must be flexible to respond to changes in the OE, policy, and resources.

CCMD campaign plans integrate posture, resources, requirements, subordinate campaigns, operations, activities, and investments that prepare for, deter, or mitigate identified contingencies into a unified plan of action.

The purpose of CCMD campaigns is to shape the OE, deter aggressors, mitigate the effects of a contingency, and/or execute combat operations in support of the overarching national strategy.

Campaign Planning

Campaigns and campaign planning follow the principles of joint operations while synchronizing efforts throughout the OE with all participants. Examples include:

- **Objective.** Clear campaign objectives must be articulated and understood across the joint force. Objectives are specified to direct every military operation toward a clearly defined, decisive, and achievable goal.
- **Unity of Command.** Unity of command means all forces operate under a single commander with the requisite authority to direct all forces employed in pursuit of a common purpose.
- **Economy of Force.** Economy of force is the judicious employment and distribution of forces to achieve campaign objectives.
- **Legitimacy.** Legitimacy maintains legal and moral authority in the conduct of operations.

Campaigns are informed by strategic guidance and the requirement to be ready to execute contingency plans.

Throughout the four planning functions beginning with mission analysis within the joint planning process (JPP), the CCDR and staff develop and update the commander's critical information requirements (CCIRs). This concurrently complements assessment activities by including information requirements critical to addressing key assessment indicators, required contingency preparations, deterrent opportunities, and the critical vulnerabilities of all actors within the OE.

Conditions, Objectives, Effects, and Tasks Linkage

For CCMD campaign plans, the CCDR develops military objectives to aid in focusing the strategy and campaign plan. CCDRs' strategies establish long-range objectives to provide context for intermediate objectives. Achieving intermediate objectives sets conditions to achieve the command's objectives. The CCDR and planners update the CCMD's strategy and theater campaign plan (TCP) based on changes to national objectives, achievement of TCP objectives, and changes in the OE.

Resource-Informed Planning (Capability Assignment, Apportionment, Allocation)

CCDRs are responsible for planning, assessing, and executing their GEF- and JSCP-directed campaign plans. The CCMDs, however, receive limited budgeting and rely on the Services and the CCMD component commands to budget for and execute campaign activities. As such, the components, Joint Force Coordinator, and joint force providers must be involved during the planning process to identify resources and tools that are likely to be made available to ensure the campaign plan is executable.

Elements of a Combatant Command Campaign Plan

The CCMD campaign plan consists of all plans contained within the established theater or functional responsibilities to include contingency plans, subordinate and supporting plans, posture plans, country-specific security cooperation sections/country plans (for geographic commands), and operations in execution.

Assessing Theater and Functional Campaign Plans

Campaign plan assessments determine the progress toward accomplishing a task, creating a condition, or achieving an objective. Campaign assessments enable the CCDR and supporting organizations to refine or adapt the campaign plan and supporting plans to achieve the campaign objectives or, with SecDef approval, to adapt the GEF-directed objectives to changes in the strategic and OEs.

Risk   CCDRs and DOD's senior leaders work together to reach a common understanding of integrated risk (the strategic risk assessed at the CCMD level combined with the military risk), decide what risk is acceptable, and minimize the effects of accepted risk by establishing appropriate risk controls.

Opportunity   CCDRs need to identify opportunities they can exploit to influence the situation in a positive direction. Limited windows of opportunity may open and the CCDR must be ready to exploit these to set the conditions that will lead to successful transformation of the conflict and thus to transition.

Operational Art and Operational Design

Operational art is the cognitive approach by commanders and staffs—supported by their skill, knowledge, experience, creativity, and judgment—to develop strategies, campaigns, and operations to organize and employ military forces by integrating ends, ways, means, and risks. Operational art is inherent in all aspects of operational design.

Operational design is the conception and construction of the framework that underpins a campaign or operation and its subsequent execution. The framework is built upon an iterative process that creates a shared understanding of the OE; identifies and frames problems within that OE; and develops approaches, through the application of operational art, to resolving those problems, consistent with strategic guidance and/or policy.

The purpose of operational design and operational art is to produce an operational approach, allowing the commander to continue JPP, translating broad strategic and operational concepts into specific missions and tasks and produce an executable plan.

The Commander's Role   Commanders distinguish the unique features of their current situations to enable development of innovative or adaptive solutions. They understand that each situation requires a solution tailored to the context of the problem. Through the use of operational design and the application of operational art, commanders develop innovative, adaptive alternatives to solve complex challenges.

Methodology

The general methodology in operational design is:
- Understand the strategic direction and guidance.
- Understand the strategic environment (policies, diplomacy, and politics).
- Understand the OE.
- Define the problem.
- Identify assumptions needed to continue planning (strategic and operational assumptions).
- Develop options (the operational approach).
- Identify decisions and decision points (external to the organization).
- Refine the operational approach(es).
- Develop planning guidance.

Elements of Operational Design

Termination. Termination criteria describe the conditions that must exist in the OE at the cessation of military operations.

Military End State. Military end state is the set of required conditions that defines achievement of all military objectives. It normally represents a point in time and/or circumstances beyond which the President does not require the military instrument of national power as the primary means to achieve remaining national objectives.

Objectives. An objective is clearly defined, decisive, and attainable. Objectives and their supporting effects provide the basis for identifying tasks to be accomplished.

Effects. An effect is a physical and/or behavioral state of a system that results from an action, a set of actions, or another effect.

Center of Gravity (COG). A COG is a source of power that provides moral or physical strength, freedom of action, or will to act.

Decisive Points. A decisive point is a geographic place, specific key event, critical factor, or function that, when acted upon, allows a commander to gain a marked advantage over an enemy or contributes materially to achieving success.

Line of Operation (LOO) and Line of Effort (LOE):

- **LOOs.** A LOO defines the interior or exterior orientation of the force in relation to the enemy or that connects actions on nodes and/or decisive points related in time and space to an objective(s).
- **LOEs.** A LOE links multiple tasks and missions using the logic of purpose—cause and effect—to focus efforts toward establishing operational and strategic conditions.

Direct and Indirect Approach. The approach is the manner in which a commander contends with a COG.

Anticipation. During execution, JFCs should remain alert for the unexpected and for opportunities to exploit the situation.

Operational Reach. Operational reach is the distance and duration across which a joint force can successfully employ military capabilities.

Culmination. Culmination is that point in time and/or space at which the operation can no longer maintain momentum.

Arranging Operations. Commanders must determine the best arrangement of joint force and component operations to conduct the assigned tasks and joint force mission.

Operational Pause. Operational pauses may be required when a major operation may be reaching the end of its sustainability.

Forces and Functions. Typically, JFCs structure operations to attack both enemy forces and functions concurrently to create the greatest possible friction between friendly and enemy forces and capabilities.

Phasing

A phase can be characterized by the focus that is placed on it. Phases are distinct in time, space, and/or purpose from one another, but must be planned in support of each other and should represent a natural progression and subdivision of the campaign or operation. Each phase should have a set of starting conditions that define the start of the phase and ending conditions that define the end of the phase. The ending conditions of one phase are the starting conditions for the next phase.

Transitions

Transitions between phases are planned as distinct shifts in focus by the joint force, often accompanied by changes in command or support relationships. The activities that predominate during a given phase, however, rarely align with neatly definable breakpoints.

Joint Planning Process

JPP is an orderly, analytical set of logical steps to frame a problem; examine a mission; develop, analyze, and compare alternative COAs; select the best COA; and produce a plan or order. The application of operational design provides the conceptual basis for structuring campaigns and operations.

The JPP seven-step process aligns with the four APEX planning functions. The first two JPP steps (planning initiation and mission analysis) take place during the APEX strategic guidance planning function. The next four JPP steps (COA development, COA analysis and wargaming, COA comparison, and COA approval) align under the APEX concept development planning function. The final JPP step (plan or order development) occurs during the APEX plan development planning function.

Planning Initiation (Step 1)

Joint planning begins when an appropriate authority recognizes potential for military capability to be employed in support of national objectives or in response to a potential or actual crisis. **At the strategic level, that authority—the President, SecDef, or CJCS—initiates planning by deciding to develop military options.** Presidential directives, NSS, UCP, GEF, JSCP, and related strategic guidance documents (e.g., strategic guidance statements) serve as the primary guidance to begin planning.

Mission Analysis (Step 2)

The CCDR and staff analyzes the strategic direction and derives the restated mission statement for the commander's approval, which allows subordinate and supporting commanders to begin their own estimates and planning efforts for higher headquarters' concurrence. **The joint force's mission is the task or set of tasks,** together with the purpose, that clearly indicates the action to be taken and the reason for doing so.

Course of Action (COA) Development (Step 3)

COA is a potential way (solution, method) to accomplish the assigned mission. The staff develops COAs to provide unique options to the commander, all oriented on accomplishing the military end state. A good COA accomplishes the mission within the commander's guidance, provides flexibility to meet unforeseen events during execution, and positions the joint force for future operations. It also gives components the maximum latitude for initiative.

COA Analysis and Wargaming (Step 4)

COA analysis is the process of closely examining potential COAs to reveal details that will allow the commander and staff to tentatively identify COAs that are valid and identify the advantages and disadvantages of each proposed friendly COA. The commander and staff analyze each COA separately according to the commander's guidance.

COA wargaming is a conscious attempt to visualize the flow of the operation, given joint force strengths and dispositions, adversary capabilities and possible COAs, the OA, and other aspects of the OE. Each critical event within a proposed COA should be wargamed based upon time available using the action, reaction, and counteraction method of friendly and/or opposing force interaction.

COA Comparison (Step 5)

COA comparison is a subjective process whereby COAs are considered independently and evaluated/compared against a set of criteria that are established by the staff and commander. The objective is to identify and recommend the COA that has the highest probability of accomplishing the mission.

COA Approval (Step 6)

In this JPP step, the staff briefs the commander on the COA comparison and the analysis and wargaming results, including a review of important supporting information. The staff determines the preferred COA to recommend to the commander.

The commander, upon receiving the staff's recommendation, combines personal analysis with the staff recommendation, resulting in a selected COA.

Plan or Order Development
(Step 7)

Planning results in a plan that is documented in the format of a plan or an order. If execution is imminent or in progress, the plan is typically documented in the format of an order. During plan or order development, the commander and staff, in collaboration with subordinate and supporting components and organizations, expand the approved COA into a detailed plan or OPORD by refining the initial CONOPS associated with the approved COA. **The CONOPS is the centerpiece of the plan or OPORD.**

Operation Assessment

Commanders maintain a personal sense of the progress of the operation or campaign, shaped by conversations with senior and subordinate commanders, key leader engagements, and battlefield circulation. Operation assessment complements the commander's awareness by methodically identifying changes in the OE, identifying and analyzing risks and opportunities, and formally providing recommendations to improve progress towards mission accomplishment. Assessment should be integrated into the organization's planning (beginning in the plan initiation step) and operations battle rhythm to best support the commander's decision cycle.

Assessment analysis and products should identify where the CCMD's ways and means are sufficient to attain their ends, where they are not and why not, and support recommendations to adapt or modify the campaign plan or its components.

Tenets of Operation
Assessment

Commander Centricity. The assessment plan should focus on the information and intelligence that directly support the commander's decision making.

Subordinate Commander Involvement. Assessments are more effective when used to support conversations between commanders at different echelons.

Integration. Operation assessment is the responsibility of commanders, planners, and operators at every level and not the sole work of an individual advisor, committee, or assessment entity.

Integration into the Planning Process and Battle Rhythm. To deliver information at the right time, the operation assessment should be synchronized with the commander's decision cycle.

Integration of External Sources of Information. Operation assessment should allow the commander and staff to integrate information that updates the understanding of the OE in order to plan more effective operations.

Credibility and Transparency. As much as possible, sources and assessment results should be unbiased. All methods used, and limitations in the collection of information and any assumptions used to link evidence to conclusions, should be clearly described in the assessment report.

Continuous Operation Assessment. While an operation assessment product may be developed on a specific schedule, assessment is continuous in any operation.

Staff Organization for Operation Assessment

The commander or chief of staff (COS) should identify the director or staff entity responsible for the collective assessment effort in order to synchronize activities, achieve unity of effort, avoid duplication of effort, and clarify assessment roles and responsibilities across the staff.

Within typical staff organizations, there are three basic locations where the responsible element could reside:
- **Special Staff Section.** In this approach, the assessment element reports directly to the commander, via the COS or deputy commander.
- **Separate Staff Section.** In this approach, the assessment element is its own staff section, akin to plans, operations, intelligence, logistics, and communications.
- **Integrated in Another Staff Section.** In this approach, the assessment element is typically integrated into the operations or plans sections, and the assessment chief reports to the plans chief or the operations chief.

Operation Assessment Process

Every mission and OE has its own unique challenges, making every assessment unique. The following steps can help guide the development of an effective assessment plan and assessment performance during execution.

- **Step 1—Develop the Operation Assessment Approach**
- **Step 2—Develop Operation Assessment Plan**
- **Step 3—Collect Information and Intelligence**
- **Step 4—Analyze Information and Intelligence**
- **Step 5—Communicate Feedback and Recommendations**
- **Step 6—Adapt Plans or Operations/Campaigns**

Linking Effects, Objectives, and End States to Tasks Through Indicators

As the staff develops the desired effects, objectives, and end states during planning, they should concurrently identify the specific pieces of information needed to infer changes in the OE supporting them. These pieces of information are commonly referred to as indicators.

Guidelines for Indicator Development

Indicators should be **relevant, observable or collectable, responsive,** and **resourced.**

Relevant. Indicators should be relevant to a desired effect, objective, or end state within the plan or order. A valid indicator bears a direct relationship to the desired effect, objective, or end state and accurately signifies the anticipated or actual status of something about the effect, objective, or end state that must be known.

Observable and Collectable. Indicators must be observable (and therefore collectable) such that changes can be detected and measured or evaluated. The staff should make note of indicators that are relevant but not collectable and report them to the commander.

Responsive. Indicators should signify changes in the OE timely enough to enable effective response by the staff and timely decisions by the commander. Assessors must consider an indicator's responsiveness to stimulus in the OE.

Resourced. The collection of indicators should be adequately resourced so the command and subordinate units can obtain the required information without excessive effort or cost.

Linking Effects, Objectives, and End States to Tasks Through Indicators

Ensuring effects, objectives, and end states are linked to tasks through carefully selected measures of performance (MOPs) and measures of effectiveness (MOEs) is essential to the analytical rigor of an assessment framework. Establishing strong, cogent links between tasks and effects, objectives, and end states through MOPs and MOEs facilitates the transparency and clarity of the assessment approach. Additionally, links between tasks and effects, objectives, and end states assist in mapping the plan's strategy to actual activities and conditions in the OE and subsequently to desired effects, objectives, and end states.

Transition to Execution

Types of Transition

There are three possible conditions for transitioning planning to execution:

- **Contingency Plan Execution.** Contingency plans are planned in advance to typically address an anticipated crisis. If there is an approved contingency plan that closely resembles the emergent scenario, that plan can be refined or adapted as necessary and executed. The APEX execution functions are used for all plans.
- **Crisis Planning to Execution.** Crisis planning is conducted when an emergent situation arises. The planning team will analyze approved contingency plans with like scenarios to determine if an existing plan applies. If a contingency plan is appropriate to the situation, it may be executed through an OPORD or fragmentary order. In a crisis, planning usually transitions rapidly to execution, so there is limited deviation between the plan and initial execution.
- **Campaign Plan Execution.** Activities within campaign plans are in constant execution. Planning is conducted based upon assumed forces and resources. Upon a decision to execute, these assumptions are replaced by the facts of actual available forces and resources. Disparities between planning assumptions and the actual OE conditions at execution will drive refinement or adaption of the plan or order.

Transition Process

The transition from plan to execution should consider the following points. These are not meant to be exclusive and may be conducted simultaneously:

- Update environmental frame and intelligence analysis.
- Identify any changes to strategic direction or guidance.
- Identify forces and resources, to include transportation.
- Identify decision points and CCIRs to aid in decision making.
- **Confirm Authorities for Execution.** Request and receive President or SecDef authority to conduct military operations.
- **Direct Execution.** The Joint Staff, on behalf of the CJCS, prepares orders for the President or SecDef to authorize the execution of a plan or order.
- **Impact on Other Operations.** As the plan transitions to execution, the commander and staff synchronize that operation with the rest of the CCMD's theater (or functional) campaign.

CONCLUSION

This publication reflects current doctrine for conducting joint, interagency, and multinational planning activities and, as a keystone publication, provides the core of joint doctrine for joint planning across the range of military operations.

CHAPTER I
JOINT PLANNING

"The tactical result of an engagement forms the base for new strategic decisions because victory or defeat in a battle changes the situation to such a degree that no human acumen is able to see beyond the first battle. . . . Therefore no plan of operations extends with any certainty beyond the first contact with the main hostile force."

Helmuth von Moltke the Elder, *On Strategy*

1. Overview

a. Joint planning is the deliberate process of determining how (the **ways**) to use military capabilities (the **means**) in time and space to achieve objectives (the **ends**) while considering the associated **risks.** Ideally, planning begins with specified national strategic objectives and military end states to provide a unifying purpose around which actions and resources are focused. The joint planning and execution community (JPEC) conducts joint planning to understand the strategic and operational environment (OE) and determines the best method for employing the Department of Defense's (DOD's) existing capabilities to achieve national objectives. Joint planning identifies military options the President can integrate with other instruments of national power (diplomatic, economic, informational) to achieve those national objectives. In the process, joint planning identifies likely benefits, costs, and risks associated with proposed military options. In the absence of specified national objectives and military end states, combatant commanders (CCDRs) may propose objectives and military end states for the President's and/or the Secretary of Defense's (SecDef's) consideration before beginning detailed planning. The Chairman of the Joint Chiefs of Staff (CJCS), as the principal military advisor to the President and SecDef, may offer military advice on the proposed objectives and military end states as a part of this process.

b. The strategic environment is uncertain, complex, and changes rapidly. While the nature of war has not changed, the character of warfare has evolved. Military operations will increasingly operate in a **transregional, multi-domain, and multi-functional** (TMM) environment. TMM operations will cut across multiple combatant commands (CCMDs) and across land, maritime, air, space, and cyberspace. Effective planning provides leadership with options that offer the highest probability for success at acceptable risk and enables the efficient use of limited resources, including time, to achieve objectives in this global environment. When specific objectives are not identified, planning identifies options with likely outcomes and risks to enable leaders at all levels to make informed decisions, without unnecessary expenditure of resources.

c. At the CCMD level, joint planning serves two critical purposes.

(1) At the strategic level, joint planning provides the President and SecDef options, based on best military advice, on use of the military in addressing national interests

and achieving the objectives in the *National Security Strategy* (NSS) and Defense Strategy Review (DSR).

(2) At the operational level, once strategic guidance is given, planning translates this guidance into specific activities aimed at achieving strategic and operational-level objectives and attaining the military end state. This level of planning ties the training, mobilization, deployment, employment, sustainment, redeployment, and demobilization of joint forces to the achievement of military objectives that contribute to the achievement of national security objectives in the service of enduring national interests.

2. Principles of Planning

a. **Focuses on the End State.** Joint planning is end state oriented: plans and actions positively contribute to achieving national objectives. Planning begins by identifying the desired national and military end states. The commander and staff derive their understanding of those end states by evaluating the strategic guidance, their analysis of the OE, and coordination with senior leadership. Joint planners must ensure plans are consistent with national priorities and are directed toward achieving national objectives. Planning must also determine and articulate the correct problem set to which military effort might be applied. The CCDR and staff work with DOD leadership in this effort. The CCDR, staff, and SecDef (or designated representative) likely view the problem from differing perspectives. Examining and discussing these perspectives is essential since a directed military end state or objective may not necessarily result in the expected strategic end state as envisioned by policymakers. Commanders and the Services, with their staffs, must identify and discuss with DOD leaders gaps between the directed military end states, the capabilities and limitations of employing the military, and the desired national end states and objectives.

b. **Globally Integrated and Coordinated.** Planning considers that operations take place throughout the OE irrespective of geographic, political, or domain boundaries. Planning, therefore, must look across CCMD, Service, and even DOD or US boundaries to ensure effective support for national objectives.

(1) Many of the challenges faced by the US transcend geographic boundaries and DOD-defined domains. Planning needs to include the broader impact of US and adversary operations and how they act, react, and interact across CCMD functional and geographic boundaries. Integrated planning coordinates resources, timelines, decision points, and authorities across CCMD functional areas and areas of responsibility (AORs) to attain strategic end states. Integrated planning produces a shared understanding of the OE, required decisions, resource prioritization, and risk across the CCMDs. CCDRs, joint force commanders (JFCs), and component commanders need to involve all associated commands and agencies within DOD in their plans and planning efforts. Moreover, planning efforts must be coordinated with other United States Government (USG) department and agency stakeholders in the execution of the plan to assure unity of effort across the whole-of-government. The integrated planning process is the way the joint force will address complex challenges that span multiple CCMD AORs and functional responsibilities.

(2) Military forces alone cannot achieve national objectives. Joint forces must effectively coordinate with USG departments and agencies, allied and partner nations, nongovernmental organizations (NGOs), international organizations, commercial entities (contractors), and local and regional stakeholders. These networks of forces and partners will form, evolve, dissolve, and reform in different arrangements in time, space, and purpose to best meet the needs of the operation or campaign. JFCs and staffs should consider how to involve interagency and multinational partners, relevant international organizations and NGOs, and the private sector in the planning process; how to coordinate and synchronize joint force actions with the operations of these organizations; and the military actions and resources required to address international organization and NGO functions when those resources are not available, consistent with existing legal authorities. Regardless of the level of involvement during the planning process, commanders and staffs must consider the impact of these various entities on joint operations.

c. **Resource Informed.** Joint planning is resource informed and time constrained. It provides a realistic assessment of the application of forces, given current readiness, availability, location, available transportation, and speed of movement. Planning assumes that an operation will employ forces and capabilities currently available—not future capabilities or capacities.

(1) When translating strategic and CCDR guidance into joint operation plans (OPLANs) and operation orders (OPORDs), planning must begin with those resources that are likely to be available at execution and identify risk where shortfalls exist. The Adaptive Planning and Execution (APEX) enterprise provides a framework for iterative dialogue and collaborative planning to discuss the merits and risks of various military options employing joint forces. Once a decision has been made at the strategic level, joint planning facilitates development of feasible, acceptable, adequate, distinguishable, and complete courses of action (COAs). Joint planning must be agile and flexible enough to provide senior leadership with the information required for critical decision making, regardless of time constraints, while ensuring they are aware of assumptions and uncertainties in the plan as a result of a truncated analysis. The iterative nature of planning drives planners to continually refine the analysis as time permits.

(2) Planners must consider that available resources may change during plan execution. For top-level plans, this could mean identifying to DOD leadership when a plan needs to change based on actual or forecasted changes in resources (e.g., forces, ammunition, transportation, budget). Planning can identify additional resources that would reduce the risk associated with the plan, if made available. The value or intensity of the national interest will determine the resources the nation is willing to expend.

d. **Risk Informed.** Assessing and articulating risks and opportunities while identifying potential mitigation strategies are fundamental to joint planning. Planning provides decision makers an honest assessment of the costs and potential consequences of military actions. Planning identifies the impact of all assumptions whether proven valid or invalid, as well as the impact of constraints and restraints imposed on the operation. In the course of developing multiple options to attain the strategic-level end state, JFCs and their planning staffs, as well as the larger JPEC, identify and communicate shortfalls in DOD's ability to resource, execute, and sustain the military operations contained in the plan, as well as the necessary actions to act on opportunities and reduce, control, or accept risk with shared knowledge of potential consequences.

e. **Framed within the OE.** Planning requires an understanding of the OE as it exists and changes. Unlike concepts and future development, adaptive planning is based on continuous monitoring and understanding of actual conditions affecting the OE such as current friendly and adversary force postures, readiness, geopolitical conditions, and adversary perceptions. Adaptive planning accommodates changes aimed at improving probability of success or mitigating risk (e.g., additional forces, partner nation contributions, agreements, or access, basing, and overflight permission needed; preparation activities, including prepositioning). However, until those decisions are made and enacted, the starting position for any plan has to be the current OE. Planners should not assume away contentious issues or conditions in order to make the plan executable or reduce risk. Adversaries can be expected to take action to set the conditions in the theater to their advantage during peacetime or times of crisis. Such actions may challenge assumptions of US plans or ways of warfare.

f. **Informs Decision Making.** Planning, even constrained by time, identifies issues and assumptions required for planning to continue, likely resource requirements, costs and cost-benefit trade-offs, and risks associated with different COAs. Discussions on these topics enable key leaders to make informed decisions that best serve the national interests.

g. **Adaptive and Flexible.** Planning is an adaptive process. It occurs in a networked, collaborative environment that requires dialogue among senior leaders; concurrent plan development; and collaboration across strategic, operational, and tactical planning levels. Early planning guidance and frequent interaction between senior leaders and planners promotes a shared understanding of the complex operational problem, strategic and military objectives, mission, planning assumptions, considerations, risks, and other key planning factors. If clear strategic-level guidance has not been provided or disconnects emerge between direction, planning assumptions, available forces capabilities, desired objectives, and end states (often due to uncertainty in the OE), frequent dialogue becomes more important to ensure joint planners and senior leaders remain synchronized in preparing, refining, and adapting plans. This facilitates adaptive planning to produce and maintain up-to-date plans. The focus is on developing options for further planning that contain a variety of viable, flexible COAs for commanders, and in the case of top priority plans, for SecDef to consider.

3. Planning

a. Planning is the deliberate process of balancing ways, means, and risk to achieve directed objectives and attain desired end states (ends) by synchronizing and integrating the employment of the joint force. It is the art and science of interpreting direction and guidance and translating it into executable activities within imposed limitations to achieve a desired objective or attain an end state. Planning enables leaders to identify cost-benefit relationships, risks, and trade-offs to determine a preferred COA.

b. Although the four planning functions of strategic-level guidance, concept development, plan development, and plan assessment are generally sequential, they often run simultaneously in the effort to accelerate the overall planning process. Leadership may direct planning staffs to refine or adapt a plan by reentering the planning process at any of the functions. During each planning function, planning is synchronized by the JPEC through ongoing civil-military dialogue, adapting for changes in guidance and the OE. For the discussion on planning functions, see Chapter II, "Strategic Guidance and Coordination," paragraph 14, "Planning Functions."

c. **Strategy, Strategic Art, Operational Art, and Operational Planning**

(1) Strategy is a prudent idea or set of ideas for employing the instruments of national power in a synchronized and integrated fashion to achieve theater, national, and/or multinational objectives. Strategy can also be described as the art and science of determining a future state/condition (ends), conveying this to an audience, determining the operational approach (ways), and identifying the authorities and resources (time, forces, equipment, money, etc.) (means) necessary to reach the intended end state, all while managing the associated risk. Strategy should not be confused with strategic-level guidance; there are numerous strategic-level documents that make up national policy. The NSS describes the worldwide interests and objectives of the US; the national means necessary to deter aggression and the adequacy of the national resources to pursue national interests. Historically, the NSS does not address specific ways to achieve the stated objectives. SecDef and the CJCS develop separate defense and military documents that describe the ways military forces will be used in coordination with the other means to pursue national interests or support policy described in the NSS. Geographic combatant commanders (GCCs) develop a theater strategy that addresses the specific application of military resources in coordination with other instruments of national power in a geographic region. Functional combatant commanders (FCCs) develop functional strategies in support of national and GCCs' theater strategies.

(2) Strategic art is the ability to understand the strategic variable (relative to the operational area [OA]) and to conceptualize how the desired objectives set forth in strategic-level guidance can be reached through the employment of military capabilities. This also includes understanding the major international diplomatic/political and security challenges impacting on US/partner success, the potential ways that the US might employ its national means to attain desired ends, and visualizing how military operations can support and/or enable our national success. Such efforts are key to developing enduring, effective strategies for sustaining military efforts over the long term where specific military operations are required. The ability to visualize and conceptualize how strategic-level success can be achieved or supported by military means is a key foundation for the application of operational art and operational design.

(3) Operational art is the application of intuition and creative imagination by commanders and staffs. Supported by their skill, knowledge, experience, creativity, and judgment, commanders seek to understand the OE, visualize and describe the desired end state, and employ assigned resources to achieve objectives. In the planning process, many activities are best done through a scientific approach such as identifying strengths and

weaknesses of the opponent, using checklists in the planning process, and comparing the outcomes of analyses. However, conflicts and war are human constructs that rely on the art and broad knowledge of commanders and planners that are not easily categorized or countered.

(4) Operational planning translates the commander's concepts into executable activities, operations, and campaigns, within resource, policy, and national limitations to achieve objectives.

d. **Understanding Problems**

(1) Recognizing and defining problems are key in distinguishing between the symptoms and root causes of problems when developing strategies and plans. Before beginning work, commanders and staffs need to ask themselves "what problem are we really being asked to solve?" as it may not be the specific problem identified in the written guidance. This question begins the civilian-military dialogue at the national level and dialogue between the supported CCDR and relevant JPEC stakeholders at the theater/functional level to ensure a shared understanding of the identified issue. For example, eliminating specific threats may not resolve the underlying causes of an insurgency, and military action may exacerbate the problem rather than solve it. Identification (ID) of underlying problems informs the commanders so they can develop their campaign plans to prevent, prepare for, or mitigate contingencies.

(2) Understanding the problem highlights the importance of defining the desired objective at the beginning of the planning process. By correctly interpreting and understanding the objectives and end states, the planner may discover, for example, the proposed planning task addresses only symptoms of the problem, rather than a solution. In such a case, through civil-military discussions, other options for using the military instrument of national power may need to be identified in support of a viable solution.

See Chapter IV, "Operational Art and Operational Design," for more detailed discussion of identifying and understanding problems.

e. **Integrated Planning.** Integrated planning is used by the joint force to address complex strategic challenges that span multiple geographic CCMD AORs and functional CCMD responsibilities. Integrated planning synchronizes resources and integrates timelines, decision matrices, and authorities across CCMDs, the rest of the interagency, and multinational partners to achieve directed strategic objectives. Integrating plan development, in-progress reviews (IPRs), and assessment provides national leadership a holistic understanding of how a particular conflict could realistically develop, options for response, and how operations by one CCMD could affect the broader OE across the globe.

See the Chairman of the Joint Chiefs of Staff Instruction (CJCSI) 3110.01, (U) 2015 Joint Strategic Capabilities Plan (JSCP), for further information on problem set integrated planning requirements.

4. Strategic, Theater, and Functional Planning

a. CCDRs use strategic guidance and direction to prepare command strategies, focused on their command's specific capabilities and missions to link national strategic guidance to theater or functional strategies and joint operations. The command strategy, like national strategy, identifies broad, long-range objectives the command aims to achieve as their contribution toward national security. The command strategy provides the link between national strategic guidance and joint planning.

(1) **CCMD Campaign Plans.** CCMD campaign plans, also known as theater campaign plans (TCPs) and functional campaign plans (FCPs), implement the military portion of national policy and defense strategy by identifying those actions the CCMDs will conduct on a daily basis. Designated campaign plans (including ongoing operations, security cooperation activities, intelligence activities, exercises, and other shaping or preventive activities) direct the activities the command will do to shape the OE and prepare for, mitigate, or deter crises on a daily basis. CCDRs identify the resources assigned and allocated to the CCMDs, prioritize objectives, and commit those resources to shape the OE and support the national strategic objectives. CCDRs evaluate the commitment of resources and make recommendations to civilian leadership on future resources and national efforts associated with executing the command's missions.

(2) **Contingency Planning.** CCDRs are directed in the *Guidance for Employment of the Force* (GEF) and *Joint Strategic Campaign Plan* (JSCP) to prepare for specific contingencies. So simultaneously, the CCDRs direct their staffs to conduct planning to address these contingencies within their region or functional area. CCDRs may also identify additional contingencies the command should prepare for through an analysis of the AOR or functional area. As a part of contingency planning, CCDRs backward plan to ensure their campaign plans address issues in the OE.

(a) Since contingency planning is based on hypothetical situations, it relies on assumptions to fill in gaps. Although contingency planning and associated end states are GEF-directed, specific conditions affecting COAs remain uncertain, making it difficult to identify specific decisions for events that have not yet occurred in a dynamic OE. CCDRs may be asked to provide multiple options to the civilian and military leadership so they can better understand how their decisions (to include timing of those decisions) can impact an operation.

(b) Contingency plans are branch plans to the CCMD campaign plans. CCDRs include the operations, activities, and investments considered critical to contingency preparation within the CCMD campaign plan to reduce the likelihood of contingency plan execution by preventing or deterring the conditions and actions leading to crises.

b. As allies, partners, competitors, and threats do not restrict their operations by US CCMD boundaries, CCDRs and their planners must integrate their plans with other CCDRs to ensure unified actions in support of national, strategic, theater, and operational objectives. Integrated planning also synchronizes resources and integrates timelines, decision points, and authorities across multiple CCMDs to achieve GEF-directed campaign objectives and attain contingency end states.

c. **Support Relationships.** Since support at the joint level is a command relationship, SecDef may identify, or CCDRs may request, designation of support relationships through an establishing directive.

(1) **Supported Commander.** The supported commander designates and prioritizes objectives, timing, and duration of the supporting action. The supported commander ensures supporting commanders understand the operational approach and the support requirements of the plan. If required, SecDef will adjudicate competing demands for resources (e.g., high demand/low density assets) when there are simultaneous requirements amongst multiple supported CCDRs.

(2) **Supporting Commander(s).** The supporting commander determines the forces, tactics, methods, procedures, and communications to be employed in providing support. The supporting commander advises and coordinates with the supported commander on matters concerning the employment and limitations (e.g., logistics) of required support, assists in planning for the integration of support into the supported commander's effort, and ensures support requirements are appropriately communicated throughout the supporting commander's organization. Identifying issues early in the planning process improves the supported commander's COA development, provides a better understanding of potential risk factors, and improves their ability to react to changing environments. The supporting commander ascertains the needs of the supported force and takes action to fulfill them, within existing capabilities, consistent with priorities and requirements of other assigned tasks. When the supporting commander cannot fulfill the needs of the supported commander, the establishing authority will be notified by either the supported or supporting commanders.

See Joint Publication (JP) 1, Doctrine for the Armed Forces of the United States, *for more information on support relationships.*

d. **Global Missions.** CCDRs can be tasked to address missions that cross geographic CCMD boundaries. CCDRs tasked with global missions provide planning and assessment expertise to identify tasks and missions other CCMDs (supporting commands) must perform to ensure success of global missions. Commands include supporting tasks as part of their campaign and contingency planning and coordinate to ensure assessments are complete. CCDRs with global responsibilities will also use the integrated planning process to provide an assessment of risk from the global, cross-AOR, perspective to ensure the military advice provided to the President and SecDef includes these considerations. Chapter III, "Strategy and Campaign Development," discusses this in more detail.

(1) At the operational level, CCDRs identify, prioritize, and sequence intermediate objectives that support the achievement of the national-level objectives. Intermediate objectives serve as waypoints against which the CCMD can measure success in attaining GEF-directed and national strategic objectives, and represent multiple actions that occur between initiation of a CCMD campaign and the ultimate achievement of

campaign objectives. Intermediate objectives should be discrete, identifiable, measurable, and achievable.

(2) At the tactical level, forces are arranged and employed to achieve a specific immediate task or mission. Although specific task accomplishment at the tactical level may not directly achieve the operational or strategic objective, the cumulative effects of the tactical events should achieve those objectives. Throughout this publication, the term "effects" is intended to mean both desired and undesired effects unless otherwise specified.

See JP 1, Doctrine for the Armed Forces of the United States, *for discussion on the levels of warfare.*

5. Strategy, Plans, Operations, and Assessments Cycle

a. Strategy, plans, operations, and assessments are inexorably intertwined. Plans translate the strategy into operations with the expectation that successful operations achieve the desired strategic objectives. Similarly, the effects of operations, successful or otherwise, change the operational and strategic environment, requiring constant evaluation of the strategic-level objectives to ensure they are still relevant and feasible. Joint forces, through their assessments, identify when their actions begin to negatively affect the OE and change their operations and activities to ensure better alignment between the actions and objectives.

b. Throughout planning and execution, commanders and staffs constantly assess conditions or effects to identify whether changes in the OE support national strategic interests. In developing the commander's information requirements, the commander and staff identify key elements of the OE as indicators for either success or failure to ensure the strategy remains on track. As necessary, the commander updates the command's strategy to reflect the changed OE and ensure continued coherence with national policy. Simultaneously, the commander also updates operations as needed to reflect the changed OE and updated strategy.

c. Since operations are not conducted in a closed system, the commander and staff must maintain iterative dialog with the JPEC stakeholders throughout planning and execution to address causal factors effectively in the OE and other affected geographical CCMDs. Causality is difficult to prove, as other actors with their own agenda affect the OE as well, changes could be due to the other actions or a combination of external and internal actions. Further, the chosen approach could affect the OE in a manner counter to the desired objective. In these instances, reframing the problem may be required as a result of the assessment and feedback process. This OE-wide assessment process, which embraces contributing CCMDs and other organizations, facilitates keeping the commander's strategic estimate updated to favorably influence strategy, planning, and execution.

6. Shared Understanding

a. **Civilian-Military Dialogue.** Strategy and joint planning occur within APEX, the department-level enterprise of policies, processes, procedures, and reporting structures supported by communications and information technology used by the JPEC to plan and execute joint operations. A focus of APEX is the interaction between senior DOD civilian leadership, CCDRs, and CJCS, which helps the President and SecDef decide when, where, and how to employ US military forces and resources. The interactive, iterative, and collaborative process within APEX guides the way in which planning and execution occurs throughout the Armed Forces of the United States. APEX provides SecDef and the President a range of military options, with associated resource requirements and risk assessments, to address identified threats and opportunities.

b. **Bridging Perspectives**

(1) Adaptive planning provides a range of options at the operational and strategic levels. The dynamics and uncertainty inherent in the strategic environment compel policymakers to retain maximum flexibility and thus their guidance will tend to be more general than military planners desire. This is driven by insufficient information, uncertainty about future resources, and developing political situations. There are advantages in initial general objectives as ends, ways, or means may need to change as the operation unfolds.

(2) CCDRs should identify how the activities and events planned as part of the campaign fulfill the objectives established by the civilian leadership. In cases where the objective is poorly defined, military leaders should request further clarity. Discussion with military leaders informs and aids the civilian policymakers in formulating their policy. This discussion should include how the command will assess the impact of the campaign activities and the opportunities and risks associated with execution, delay, or cancellation of those activities. The discussion should also cover how the campaign could establish conditions to prevent, prepare for, or mitigate contingencies.

c. **Identifying Desired End States and Objectives**

(1) **End State.** An end state defines achievement for all objectives. The **military end state** normally represents a period in time or set of conditions beyond which the President does not require the military instrument of national power as the primary means to achieve remaining national objectives. Commanders and planners constantly assess the stated end state against the OE, resources, or policy.

(2) **Objectives.** Objectives are clearly defined, decisive, and attainable goals toward which every operation is directed. These are short- to mid-range goals that are specific, measurable, achievable, relevant, and time-bound. Objectives are used as markers, during the execution and assessment of the strategy and aid in developing decision points. CCDRs should identify intermediate objectives as steps to aid in assessing progress toward the longer-range objectives established by the GEF or JSCP. As intermediate objectives are achieved, commanders and their staffs reassess their vision of the end state, their progress toward the longer-range objectives, and the need to change or alter the objectives or methods.

d. **Providing Options, Aligning Resources**

(1) The joint planning process (JPP) is a proven problem-solving technique designed for military planning. The planning staff typically uses JPP to conduct detailed planning to fully develop options, identify resources, and identify and mitigate risk. Planners develop the concept of operations (CONOPS), force plans, deployment plans, and supporting plans that contain multiple COAs in order to provide the flexibility to adapt to changing conditions and remain consistent with the JFC's intent and present acceptable options to civilian decision makers.

(2) CCDRs provide options for the use of the military in conjunction with other instruments of national power. Further planning enables them to develop COAs that identify costs and risks associated with the options, a timeline, required resources and capabilities, likely costs (including casualties), and probability of success or failure of the military objectives in contributing to the desired national strategic objectives.

(3) Shared understanding includes leaders (both civilian and military) identifying expected contributions from other USG departments and agencies and how they could affect military and strategic success. Interagency planning should ensure these expectations are both shared by all agencies and are realistic, based on agency capabilities and capacity.

(4) Early in the planning process, civilian and military leaders need to identify partner nations' contributions, requirements, and impacts. They must identify who will open discussions and the timing.

Chapter V, "Joint Planning Process," discusses JPP in more detail.

7. **Risk Identification and Mitigation**

   a. **Identifying Risk**

      (1) Risk assessment is initially conducted during mission analysis and is updated throughout the planning process. During planning, assumptions, which are logical, realistic, and essential for planning to continue, are used in the absence of facts. Assumptions are reviewed continuously to determine their continued validity. An assumption used in planning may subsequently cause the development of a branch plan. When sufficient information is received to invalidate an assumption, it may create the need to make changes to the plan or develop a new COA or plan.

      (2) Along with hazard and threat analysis, force requirements are determined, and shortfall ID is performed throughout the plan development process. The supported commander continuously identifies limiting factors, capability shortfalls, opportunities, and associated risks as plan development progresses. Where possible, the supported commander resolves the shortfalls through planning adjustments and coordination with supporting commanders and subordinate commanders and Services. If there is a reasonable expectation that the required resources will not become available, then the CCMD must develop an alternative approach within the means available or can be reasonably expected to become available. To identify shortfalls, the CCMD makes assumptions as to the sourcing feasibility of force requirements. CCMDs are encouraged to solicit the advice of

the Services, other CCMDs, joint force providers (JFPs), Joint Staff J-35 [Joint Force Coordinator], joint functional managers (JFMs) (as applicable), and other force providers (FPs) in identifying preferred forces. If the shortfalls and necessary controls and countermeasures cannot be reconciled or the resources provided are inadequate to perform the assigned task, the commander reports these limiting factors and assessment of the associated risk to the CJCS. The CJCS and Joint Chiefs of Staff (JCS) consider shortfalls and limiting factors reported by the supported commander and coordinate resolution. However, the continued development of assigned plans is not delayed pending the resolution of shortfalls, and the commander remains responsible for developing strategies for mitigating the risk. The JFPs work collaboratively with the Services (via their assigned Service components) and other CCDRs to provide recommended sourcing solutions to the Joint Staff (JS).

b. **Mitigating Risk.** As part of the planning process, and in discussions with senior leaders, planners and CCDRs identify possible methods to mitigate the risk associated with any plan. Some methods of mitigating risk are:

(1) **Reducing Likelihood of Occurrence.** Mitigate risk by decreasing the likelihood that events that can negatively affect our efforts will occur. Examples include the inclusion of protective safety measures (e.g., mandating the use of malaria prophylaxis in high risk areas), funding installation resiliency efforts (e.g., redundancy in critical infrastructure and systems at forward locations), and avoiding a potential hazard (e.g., using proven low-water crossings rather than untested bridges).

(2) **Reducing Cost of Occurrence.** Mitigate risk by decreasing the potential negative effect of these events if they were to occur. Examples include the inclusion of reactive safety measures (e.g., placing a corpsman/medic with an infantry platoon) and dispersion (e.g., placing capabilities at multiple locations so that an attack at one will affect only capacity).

(3) **Nonorganic Support.** Mitigate risk by use of contracted support or host-nation support (HNS) to address shortfalls in forces, limitations associated with strategic lift, and to enable the deployment of combat forces in lieu of combat service support forces.

c. **Residual Risk.** Regardless of the efforts to mitigate risk, some level of risk will remain. This should be identified to senior leaders so there is a common understanding of the decisions required and their potential effects. One of the most important roles of a commander is acknowledging and accepting these residual risks prior to executing a mission.

d. **Risk Discussion.** Commanders must include a discussion of risk in their interaction with DOD senior leaders.

(1) This discussion must be in discrete, concrete terms that enable and support decision making. Identifying risk as "high" does not support decision making, as it provides no context for a decision in relation to other strategic choices. Not all elements of risk can be quantified; analytic and modeling outputs are not always accurate. However,

by stating that "in our analysis, the mission will take six months versus two months," or "we expect casualties to increase from x to y," senior leaders are better informed of the relative difference between differing COAs.

(2) At the strategic level, CCDRs should provide feedback to senior civilian leaders and stakeholders on the implications and risk associated at the strategic level (e.g., impact on the US public, allies, adversaries, US objectives, and future US status). Although the CCDRs provide strictly the military position, in many cases, they have had a broader exposure to the implications and impact of the employment of the military both at home and abroad and the discussion may identify issues not previously noted.

(3) The methods used to mitigate risk in planning should also be identified in the discussion. DOD leadership and the President may have differing opinions on methods to mitigate risk and may identify options not previously available to the CCDR and planners.

8. Assessment

a. Plans are continually assessed by CCDRs and reviewed by the JPEC and senior DOD leadership. Assessment is a continuous process that measures the overall effectiveness of employing joint force capabilities during military operations and the expected effectiveness of plans against contingencies as the OE changes. Operation assessment is a continuous process that supports decision making by measuring the progress toward accomplishing a task, creating an effect, achieving an objective, or attaining a military end state. The purpose of assessment is to integrate relevant, reliable feedback into planning and execution, thus supporting the commander's decision making regarding plan development, adaptation, and refinement, as well as adjustment of operations during execution. A secondary purpose is to inform civil-military leadership to support geopolitical and resource decision making.

b. Assessment involves monitoring and analyzing changes in the OE, determining the most likely potential causes for those changes, identifying opportunities and risks, and providing recommendations for improving operation or campaign performance to achieve objectives. The assessment of a plan or campaign links operations, actions, and investments with desired objectives and end states. Integrating assessment planning throughout plan development and post-approval refinement and adaptation helps keep the plan relevant and ready for transition to execution.

c. Commanders are the central driver for assessments as the ultimate stakeholders in the success of their command's activities. The commander must continually monitor the OE and assess the progress toward desired objectives, as well as the effectiveness of the operation to attain the end state(s). Commanders cannot accomplish this through "gut instinct" alone. Commanders are assisted by a collective assessment effort from their staffs and subordinate commanders, along with interagency and multinational partners and other stakeholders. Assessments allow commanders to direct adjustments to plans and orders, thus ensuring the operation remains focused on accomplishing the mission. Operation assessment is applicable across the range of military operations and offers perspective and insight, providing the opportunity for self-correction, adaptation, and thoughtful results-oriented learning. However, assessment mechanisms and the assessment processes may differ at the tactical, operational, theater, global, and other strategic levels dependent upon the commander's pace of decision making and the availability of OE analysis capabilities.

d. Assessment is a continuous operational activity that spans both planning and execution functions.

(1) Effective operation assessments link the employment of forces and resources to intelligence analysis of the OE. Using an operation assessment framework helps to organize and analyze the data and communicate recommendations to the commander in accordance with the assessment plan. This enables the commander to identify the information and intelligence necessary to conduct the operations' assessment and to build those processes into the plan so the staff and commander can monitor progress or regression to implement necessary changes during execution. The assessment framework and assessment plan is a reflection of the plan and linkages of elements within the plan (objectives/end states linked to military objectives linked to effects/conditions linked to key tasks).

(2) Throughout JPP and execution, assessment helps commands analyze changes in the OE, changes in strategic guidance, and other challenges facing the joint force, in order to adapt and update plans and orders to effectively achieve objectives. Changes in the OE are the result of constant interaction between adversary, friendly, and neutral elements. Changes include random and unpredictable events, or friction, that complicate or challenge execution of the plan. Feedback, generated from the assessment process, helps identify changes in the OE and forms the basis for learning, adaptation, and subsequent recommendations to adapt the plan and plan execution. These recommendations help the commander and staff ensure operations, actions, and investments are effective, correctly aligned with resources, focused on objectives, and contributing to the achievement of directed strategic objectives.

(3) CCDRs with global responsibilities have an assessment role. During planning, they provide planning and assessment input to integrate requirements into the plans and operations of the affected CCMDs. In execution, CCMD assessments provide an evaluation of global progress against the functional objectives to ensure the achievement of national objectives from a cross-AOR perspective.

See Chapter VI, "Operation Assessment," for additional information on planning and conducting operation assessment.

9. Agility, Initiative, and Simplicity

a. The key tenets of a plan–the commander's mission, intent, and objectives are likely to endure, subject to changes in policy and/or strategy. Operation assessment provides the means to review their validity, and reaffirm or adjust as necessary. Meanwhile, based on continuous operation assessment, the scheme of maneuver (including supporting effects and planned activities) and main effort are likely to be refreshed more frequently as the plan progresses and the command seeks to maintain the initiative.

b. The expression "failing to plan is planning to fail" may be true, but a commander will use judgment to decide how much planning is required and to what level of detail. In planning, it may be counterproductive to overthink what is inherently complex and uncertain. Chairman of the Joint Chiefs of Staff Manual (CJCSM) 3105.01, *Joint Risk Analysis,* describes some of the risks associated with reverse engineering success, based upon unrealistic assumptions of causality and predictability (including the compliance of other actors). Placing absolute faith in predetermined and closely sequenced plans is unlikely to prove successful against an agile opponent. A commander should maintain a balance between proactive planning and timely adaptation to unforeseen events as the OE changes and other relevant actors, including the adversary and competitors, adapt. Assessment-led decision making and adaptive planning are underpinned by a mindset that seeks to exploit opportunities and reverse set-backs.

c. Commanders should encourage initiative among the staff so opportunities to exploit unexpected changes in the situation are not overlooked. Recognizing how a situation is changing, identifying the implications, and exploiting opportunities as they arise are keys to success.

10. Interorganizational Planning and Coordination

Interorganizational planning and coordination is the interaction that occurs among elements of DOD; participating USG departments and agencies; state, territorial, local, and tribal agencies; foreign military forces and government departments and agencies; international organizations; NGOs; and the private sector for the purpose of accomplishing an objective. Unity of effort is the coordination, integration, and/or synchronization of the activities of these governmental and nongovernmental entities with military operations in order to achieve unified action. Successful coordination of interorganizational and multinational plans facilitates unity of effort among multiple organizations by promoting common understanding of the capabilities, limitations, and consequences of military and nonmilitary actions. It also assists with identifying common objectives and the ways in which military and civilian capabilities best complement each other to achieve these objectives.

a. **Interagency Coordination.** Interagency coordination is the interaction that occurs among USG departments and agencies, including DOD, for the purpose of accomplishing an objective. Interagency coordination forges the vital link between the US military and the other instruments of national power.

b. Achieving national strategic objectives requires effective unified action resulting in unity of effort. This is accomplished by collaboration, synchronization, and coordination of the diplomatic, informational, military, and economic instruments of national power. In such situations, military power is used in conjunction with the other instruments of national power to advance and defend US values, interests, and objectives. To accomplish this integration, the CCMDs, Services, and DOD agencies interact with non-DOD agencies and organizations to ensure mutual understanding of the capabilities, limitations, and consequences of military and nonmilitary actions, as well as the understanding of end state and termination requirements. They also identify the ways in which military and civilian capabilities best complement each other. The National Security Council (NSC) plays a key role in the integration of all instruments of national power by facilitating mutual understanding and cooperation and is responsible for overseeing the interagency planning efforts. Further, military and civilian organizations sharing information, cooperating, and striving together to accomplish a common goal is the essence of multi-organizational coordination that makes unity of effort possible. In operations involving interagency partners and other stakeholders, where the commander does not control all elements, the commander seeks cooperation and builds consensus to achieve unity of effort. Interagency and multinational consensus building is a key element to unity of effort.

c. Through all stages of planning for campaigns, contingencies, and crises, CCDRs and subordinate commanders should seek to involve relevant USG departments and agencies as directed by the GEF and other strategic guidance. CCDRs should determine early on which USG departments and agencies are the most vital as supporting or supported elements of their plans and work through the JS and the Office of the Under Secretary of Defense for Policy (OUSD[P]). Generally, interagency dialogue and coordination occurs through the IPR process and the Promote Cooperation process, led by OUSD(P) and Joint Staff J-5 [Strategic Plans and Policy], with SecDef receiving an update on the scope, scale, and substance of planning exchanges with civilian and multinational counterparts. The Promote Cooperation process specifically focuses on interagency partner input and socialization of DOD plan development.

d. Effective collaboration and coordination with interagency partners can be a critical component to successful operation and campaign activities, as well as during transitions when JFCs may operate in support of other USG departments and agencies. JFCs and their staffs must consider how the capabilities of DOD and these departments and agencies can assist each other in accomplishing the broader national strategic objectives. GCCs should coordinate directly with interagency representatives within their own command and with those in the National Capital Region through the Promote Cooperation process to obtain appropriate agreements that support their plans. This cooperation provides valuable opportunities for the command to coordinate on key issues such as overflight rights and access agreements. Coordination with NGOs should normally be done through the United States Agency for International Development (USAID) senior development advisor assigned to each geographic CCMD or through the lead federal agency for contingencies in the US.

e. The Office of the Secretary of Defense (OSD) and JS, in consultation with the Services, National Guard Bureau, and CCMDs, facilitate interagency support and coordination to support DOD plans as required. While supported GCCs are the focal points for interagency coordination in support of operations in their AORs, interagency coordination with supporting commanders is just as important. At the operational level, subordinate commanders should consider and integrate interagency capabilities into their estimates, plans, and operations.

f. The APEX enterprise facilitates interagency review of plans and appropriate annexes approved by the OUSD(P) following guidance provided in IPRs. Interagency plan reviews differ from DOD JPEC plan reviews in that inputs from non-DOD agencies are requested but not required. Additionally, non-DOD agency inputs are advisory in nature and, while a valued part of the process, do not carry veto authority. Nevertheless, provision is made for participating agencies to follow up on issues surfaced during the review in accordance with guidance from the OUSD(P).

g. **Planning and Coordination with Other Agencies.** The supported commander integrates interagency input and concerns into the joint plan. Annex V (Interagency Coordination) is one tool that can be used to collaborate planning with interagency partners. CCMDs should seek approval from OSD for full releasability of this annex to all relevant USG departments and agencies during development to ensure inputs are considered and incorporated at the earliest stage practicable. Annex V should specify the objectives, tasks, and desired level of shared situational awareness required to resolve the situation and identify the anticipated capabilities required to accomplish tasks. This common understanding enables interagency planners to more rigorously plan their efforts in concert with the military, to suggest other activities or partners that could contribute to the operation, and to better determine support requirements. The staff considers interagency participation for each phase of the operation (see Chapter IV, "Operational Art and Operational Design," for a discussion of phases).

h. **Interagency Considerations**

(1) A number of factors can complicate the coordination process, including the USG departments' or agencies' differing and sometimes conflicting policies, legal authorities, roles and responsibilities, procedures, decision-making processes, and culture. Operations may be executed by nonmilitary organizations or perhaps even NGOs with the military in support. In such instances, the understanding of military authorities, end state, and termination requirements may vary among the participants. The JFC must ensure interagency partners clearly understand military capabilities, requirements, operational limitations, liaison, and legal considerations and military planners understand the nature of the relationship and the types of support they can provide. Planners must make every effort to learn the supported organization's process, policy, and operational limitations to better identify areas where they can be of assistance. The joint force planner should also understand the supported organization's planning process (such as federal interagency operational plans, or incident command systems for crisis planning) and how those processes align with JPP. The JFC's civil-military operations center can facilitate these relationships. In the absence of a formal command structure, JFCs may be required to build consensus to achieve unity of effort. Robust liaison facilitates understanding, coordination, and mission accomplishment. Annex V to the plan or order should address all these considerations.

(2) Commanders and planners must identify the desired contributions of other agencies and organizations and communicate needs to OSD. Further, commanders and planners should integrate limitations into their planning, such as indicating where agencies cannot act. It is critical to identify and communicate risk to mission accomplishment.

Potential mitigation strategies should include COAs that do not entail the use of the military.

(3) Interagency planning (and execution) requires constant coordination to ensure agency plans, including DOD remain coordinated as guidance and the situation on the ground evolve.

(4) The President, advised by the NSC, provides strategic direction to guide the efforts of USG departments and agencies and organizations that represent all instruments of national power.

For additional information on interagency considerations, see JP 3-08, Interorganizational Cooperation.

11. Multinational Planning and Coordination

a. **General. Multinational** operations is a collective term to describe military actions conducted by forces of two or more nations. Such operations are usually undertaken within the structure of a coalition or alliance, although other possible arrangements include supervision by an international organization (such as the United Nations [UN] or Organization for Security and Cooperation in Europe). **Key to any multinational operation is unity of effort** among national and military leaders of participating nations emphasizing common objectives and shared interests as well as mutual support and respect. Agreement on clearly identified strategic and military end states for the multinational force (MNF) is essential to guide all multinational coordination, planning, and execution. Additionally, the cultivation and maintenance of personal relationships between counterparts in the participating nations are fundamental to achieving success. At times, US national interests may not be in complete agreement with those of the multinational organization or some of its individual nation states. In such situations, additional consultations and coordination will be required at the political and military levels for the establishment of a common set of operational objectives to support unity of effort among nations.

b. Joint planning will frequently be accomplished within the context of multinational planning. There is no single doctrine for multinational action, and each MNF develops its own protocols, OPLANs, concept plans (CONPLANs), and OPORDs. US planning for multinational operations should accommodate and complement such protocols and plans. JFCs must also anticipate and incorporate planning factors such as domestic and international laws, regulations, and operational limitations on the use of contributed forces, various weapons, and tactics.

(1) Joint forces should be trained and equipped for combat and noncombat operations with forces from other nations within the framework of an MNF under US or a foreign commander.

(2) MNF commanders develop multinational strategies and plans in multinational channels. Supporting US JFCs perform planning for multinational operations in US national channels. Coordination of these separate planning channels occurs at the national

level by established multinational bodies or member nations and at the theater-strategic and operational levels by JFCs, who are responsible within both channels for planning matters. US doctrine and procedures for joint planning also are conceptually applicable to multinational challenges, and the general considerations for interaction with international organizations and partner-nation organizations are similar to those for interaction with USG departments and agencies. The fundamental issues are much the same for both situations.

c. **Operational-Level Integration.** The commander of US forces dedicated to a multinational military organization integrates joint planning with multinational planning at the operational level. Normally, this will be the GCC or the subordinate JFC responsible for the geographic area within which multinational operations are to be planned and executed. These commanders always function within two chains of command during any multinational operation: the multinational chain of command and the US national chain of command. Within the multinational organizations, they command or support the designated MNF and plan, as appropriate, for multinational employment in accordance with strategic guidance emanating from multinational leadership. Within the US chain of command, they command US forces and prepare plans in response to strategic direction from the President, SecDef, and the CJCS. These tasks include developing plans to support each multinational commitment within the GCC's AOR and planning for unilateral US contingencies within the same area. In this dual capacity, the US commander coordinates multinational planning with US planning.

(1) For example, within the Asia-Pacific region, the Multinational Planning Augmentation Team, a cadre of military planners from Asia-Pacific Rim nations led by US Pacific Command, produced a MNF standard operating procedure (SOP). The intent of this MNF SOP is to increase the speed of response, interoperability, mission effectiveness, and unity of effort in MNF operations during crisis action situations. The MNF SOP will help to reduce the ad hoc nature of multinational crisis planning by establishing common "operational start points" for MNF operations and establishing SOPs for the MNF headquarters.

(2) Similarly, for North Atlantic Treaty Organization's (NATO's) operations, US and other NATO countries have developed and ratified an Allied joint doctrine hierarchy of publications that outlines the doctrine and tactics, techniques, and procedures that should be used during NATO operations. JFCs, their staffs, and subordinate forces should have access to and review and train with these publications prior to participating in NATO operations.

12. **Strategic Guidance for Multinational Operations**

a. Multinational operations start with the diplomatic efforts to create a coalition or spur an alliance into action. Discussion and coordination between potential participants initially address basic questions at the national strategic level. These senior-level discussions could involve international organizations such as the UN or NATO, existing MNFs, or individual nations. The result of these discussions should:

(1) Determine the nature and limits of the response.

(2) Determine the command structure of the response force.

(3) Determine the essential strategic guidance for the response force to include military objectives and the desired strategic and military end states.

b. In support of each MNF, a hierarchy of bilateral or multilateral bodies is established to define strategic and military end states and objectives, to develop strategies, and to coordinate strategic guidance for planning and executing multinational operations. Through dual involvement in national and multinational security processes, US national leaders integrate national and theater strategic planning with that of the MNF. Within the multinational structure, US participants work to develop objectives and strategy that complement US interests and assigned missions and tasks for participating US forces that are compatible with US capabilities. Within the US national structure, international commitments impact the development of the *National Military Strategy* (NMS) and CCDRs should adequately address relevant concerns in strategic guidance for joint planning.

c. Much of the information and guidance provided for unified action and joint operations remains applicable to multinational operations. However, commanders and staffs consider differences including, but not limited to, partners' laws, doctrine, organization, weapons, equipment, terminology, culture, politics, religion, language, and caveats on authorized military action throughout the entire operation. CCDRs and JFCs develop plans to align US forces, actions, and resources in support of the multinational plan.

d. When directed, designated US commanders participate directly with the armed forces of other nations in preparing bilateral contingency plans. Commanders and their staff assess the potential constraints, opportunities, security risks, and any additional vulnerabilities resulting from bilateral planning, and how these plans impact the ability of the US to attain its end states. Bilateral planning involves the preparation of combined, mutually developed, and approved plans governing the employment of the forces of two nations for a common contingency. Bilateral planning may be accomplished within the framework of a treaty or alliance or in the absence of such arrangements. Bilateral planning is accomplished in accordance with specific guidance provided by the President, SecDef, or CJCS and captured in a bilateral strategic guidance statement (SGS) signed by the leadership of both countries.

13. Review of Multinational Plans

US joint strategic plans or contingency plans prepared in support of multinational plans are developed, reviewed, and approved exclusively within US operational channels. They may or may not be shared in total with multinational partners. Selected portions and/or applicable planning and deployment data may be released in accordance with CJCSI 5714.01, *Policy for the Release of Joint Information.* USG representatives and commanders within each multinational organization participate in multinational planning and exchange information in mutually devised forums, documents, and plans. The formal review and approval of multinational plans is accomplished in accordance with specific procedures adopted by each multinational organization and may or may not include separate US review or approval. Multilateral contingency plans routinely require national-level US approval.

JP 3-16, Multinational Operations, *and JP 4-08,* Logistics in Support of Multinational Operations, *provide greater detail. The Multinational Planning Augmentation Team MNF SOP, available at http://community.apan.org/, provides commonly agreed upon formats and procedures that may assist with planning efforts in a multinational environment.*

Intentionally Blank

CHAPTER II
STRATEGIC GUIDANCE AND COORDINATION

> *"The higher level of grand strategy [is] that of conducting war with a far-sighted regard to the state of the peace that will follow."*
>
> **B. H. Liddell Hart, *Strategy***

1. Overview

a. This chapter introduces some of the major sources of planning guidance available to the commander and staff. It describes how strategic direction is established within the APEX enterprise and how it is implemented within the JPEC to develop military plans and orders. Finally, it discusses how to integrate other USG departments and agencies and multinational partners into overall joint planning efforts.

b. The President, SecDef, and CJCS provide their orders, intent, strategy, direction, and guidance via strategic direction to the military to pursue national interests within legal and constitutional limitations. They generally communicate strategic direction to the military through written documents, but it may be communicated by any means available. Strategic direction is contained in key documents, generally referred to as strategic guidance. Strategic direction may change rapidly in response to changing situations, whereas strategic guidance documents are typically updated cyclically and may not reflect the most current strategic direction.

SECTION A. NATIONAL AND DEPARTMENT OF DEFENSE GUIDANCE

2. Introduction

The NSC develops and recommends national security policy options for Presidential approval. The NSC is the President's principal forum for considering national security and foreign policy matters with senior national security advisors and cabinet officials. NSC decisions may be directed to any of the member departments or agencies. The President chairs the NSC. Its regular attendees (both statutory and nonstatutory) are the Vice President, Secretary of State, Secretary of the Treasury, SecDef, Secretary of Homeland Security, and Assistant to the President for National Security Affairs. CJCS is the statutory military advisor to the NSC, and the Director of National Intelligence is the intelligence advisor. For DOD, the President's decisions drive SecDef's strategic guidance, which CJCS may refine. To carry out Title 10, United States Code (USC), statutory responsibilities, the CJCS utilizes the Joint Strategic Planning System (JSPS) to provide a formal structure in aligning ends, ways, and means, and to identify opportunities and mitigate risk for the military in shaping the best assessments, advice, and direction of the Armed Forces for the President and SecDef.

3. Strategic Guidance and Direction

a. **The President.** The President provides strategic guidance through the NSS, Presidential policy directives (PPDs), executive orders, and other strategic documents in conjunction with additional guidance and refinement from the NSC. The President also signs the Unified Command Plan (UCP) and the contingency planning guidance in the SecDef-signed GEF, which are both developed by DOD.

b. **SecDef.** SecDef has authority, direction, and control over DOD. SecDef oversees the development of broad defense policy goals and priorities for the deployment, employment, and sustainment of US military forces based on the NSS. For planning, SecDef provides guidance to ensure military action supports national objectives. SecDef approves assignment and allocation of forces.

c. **Under Secretary of Defense for Policy (USD[P]).** USD(P) assists SecDef with preparing written policy guidance for the preparation of plans, reviewing plans, and assisting SecDef with other duties.

d. **CJCS.** The CJCS provides independent assessments; serves as principal military advisor to the President, SecDef, and the NSC; and assists the President and SecDef with providing unified strategic direction to the Armed Forces. In this capacity, the CJCS develops the NMS and the JSCP, which provide military implementation strategies and planning direction. The CJCS is responsible for global integration, providing advice to President of the United States and the SecDef on ongoing military operations and advising on the allocation and transfer of forces among GCCs and FCCs, as necessary, to address TMM threats. The CJCS provides additional strategic planning guidance and policy to the CCMDs and Services via CJCS directives, joint doctrine, force apportionment tables, and planning orders (PLANORDs). The CJCS also issues orders on behalf of the President or SecDef.

4. National Security Council System

The NSC system is the principal forum for interagency deliberation of national security policy issues requiring Presidential decision. In addition to NSC meetings chaired by the President, the current NSC organization includes the Principals Committee, deputies committee, and interagency policy committees. Specific issue interagency working groups support these higher-level committees. The purpose of the NSC is to develop policy recommendations with interagency consensus to the President for approval. When implemented, the policy provides strategic direction for military planning and programming.

For additional information, see PPD-1, Organization of the National Security System, *and CJCSI 5715.01,* Joint Staff Participation in Interagency Affairs.

5. National Security Strategy

a. The **NSS** is required annually by Title 50, USC, Section 3043. It is prepared by the Executive Branch of the USG for Congress and outlines the major national security concerns of the US and how the administration plans to address them using all instruments of national power. The document is often purposely general in content, and its implementation by DOD relies on elaborating direction provided in supporting documents (e.g., such as the DSR, GEF, and NMS).

b. JFCs and their staffs can derive the broad overarching policy of the US from the NSS, but must check other DOD and military sources for refined guidance as the NSS is too broad for detailed planning.

6. Department of State and the United States Agency for International Development

The Department of State (DOS) is the lead US foreign affairs agency within the Executive Branch and the lead institution for the conduct of American diplomacy. The Secretary of State is the President's principal foreign policy advisor. The Secretary of State implements the President's foreign policies worldwide through DOS and its employees. USAID is an independent federal agency that receives overall foreign policy guidance from the Secretary of State.

a. **Quadrennial Diplomacy and Development Review.** The Quadrennial Diplomacy and Development Review provides a blueprint for advancing America's interests in global security, inclusive economic growth, climate change, accountable governance, and freedom for all. As a joint effort of DOS and USAID, the review identifies major global and operational trends that constitute threats or opportunities, delineates priorities, and reforms to ensure our civilian institutions are in the strongest position to shape and respond to a rapidly changing world.

b. **DOS-USAID Joint Strategic Plan.** This DOS-USAID Joint Strategic Plan is a blueprint for investing in America's future and achieving the goals the President laid out in the NSS and those in the Quadrennial Diplomacy and Development Review. It lays out strategic goals and objectives for four years and includes key performance goals for each objective.

c. The following are key DOS/USAID planning documents that commanders and planners must consult when developing theater plans.

(1) **Joint Regional Strategies.** A joint regional strategy is a three-year regional strategy developed jointly by the regional bureaus of DOS and USAID. It identifies the priorities, goals, and areas of strategic focus within the region. Joint regional strategies provide a forward-looking and flexible framework within which bureaus and missions prioritize desired end states, supporting resources, and response to unanticipated events.

(2) **Integrated Country Strategies.** An integrated country strategy is a three-year strategy developed by a DOS country team for a particular country. It articulates a common set of USG priorities and goals by setting the mission goals and objectives through a coordinated and collaborative planning effort. It provides the basis for the development of the annual mission resource requests. The chief of mission leads the development process and has final approval authority.

(3) **Country Development Cooperation Strategy.** The country development cooperation strategy is a five-year country-level strategy that focuses on USAID-implemented assistance, including nonemergency humanitarian and transition assistance and related USG non-assistance tools.

7. Department of Defense

a. **DSR.** The DSR is legislatively mandated by Congress per Title 10, USC, Section 118, and required every four years. The DSR articulates a defense strategy consistent with the most recent NSS by defining force structure, modernization plans, and a budget plan allowing the military to successfully execute the full range of missions within that strategy for the next 20 years. The DSR flows from the NSS, informs the NMS, and provides the foundation for other DOD strategic guidance, specifically on planning, force development, and intelligence.

b. **UCP.** The UCP, signed by the President, establishes CCMD missions and CCDR responsibilities, addresses assignment of forces, delineates geographic AORs for GCCs, and specifies responsibilities for FCCs. The unified command structure identified in the UCP is flexible and changes as required to accommodate evolving US national security needs. Title 10, USC, Section 161, tasks CJCS to conduct a review of the UCP "not less often than every two years" and submit recommended changes to the President through SecDef. This document provides broad guidance that CCDRs and planners can use to derive tasks and missions during the development and modification of CCMD plans.

c. **GEF.** The GEF, signed by SecDef, and its associated Contingency Planning Guidance, signed by the President, convey the President's and the SecDef's guidance for contingency force management, security cooperation, and posture planning. The GEF translates NSS objectives into prioritized and comprehensive planning guidance for the employment of DOD forces.

(1) **Campaign Plans**

(a) CCMD campaign plans are the centerpiece of the CCMDs' planning construct and operationalize CCMD strategies. CCMD campaign plans focus the command's day-to-day activities, which include ongoing operations, military engagement, security cooperation, deterrence, and other shaping or preventive activities. CCMD campaign plans organize and align operations, activities, and investments with resources to achieve the CCMD's objectives and complement related USG efforts in the theater or functional area.

(b) **Subordinate Campaign Plans.** The CCDR or a subordinate JFC may conduct a subordinate campaign to accomplish (or contribute to) military strategic or operational objectives in support of the CCMD's TCPs and FCPs. The CCDR or subordinate JFCs develop subordinate campaign plans if their assigned missions require military operations of substantial size, complexity, and duration and cannot be accomplished within the framework of a single joint operation. These campaigns are conducted in support of the CCDR's ongoing CCMD campaign plans.

(2) **Contingency Plans.** Contingency plans are branches of TCPs or FCPs that are planned for designated threats, catastrophic events, and contingent missions without a crisis at-hand. The UCP, GEF, and JSCP guide the development of contingency plans, which address potential threats that put one or more national interest at risk in ways that

warrant military operations. Contingency plans are built to account for the possibility that campaign activities could fail to prevent aggression, preclude large-scale instability in a key state or region, or respond to a natural disaster.

(3) **Global Posture.** The GEF and JSCP provide DOD-wide global defense posture (GDP) (forces, footprint, and agreements) guidance, to include DOD's broad strategic themes for posture changes and overarching posture planning guidance, which inform the JSCP theater and functional posture planning guidance. Global posture establishes the requirement for CCDRs to submit theater posture plans (TPPs) every two years (with annual updates) to support campaign and contingency plans. Posture plans align basing and forces to ensure theater and global functional security, respond to contingency scenarios, and provide strategic flexibility.

(4) **Global Distribution.** The GEF and JSCP describe DOD's broad strategic themes for global distribution and posture that are coordinated through United States Transportation Command's (USTRANSCOM's) horizontal and vertical synchronization of global distribution planning. As a "plan of plans" the GCCs' TCPs include regional country plans, posture plans, and theater distribution plans (TDPs) that facilitate synchronization of resources, authorities, processes, and timelines in order to favorably affect conditions within the GCCs' AORs. Global distribution establishes the requirement for GCCs to submit TDPs annually to support campaign and contingency plans. Distribution plans support TCPs by interfacing with the GCCs' TPPs support to strategic lift, infrastructure, distribution enablers, agreements, policies, processes, and information systems.

For more information on posture plans, see Appendix H, "Posture Plans."

(5) **Cyberspace.** The GEF and JSCP provide campaign and integrated planning guidance for cyberspace and cyberspace operations. The potential for widespread effects across multiple functional and geographic boundaries requires US Cyber Command to synchronize operations within cyberspace. CCMDs must identify their requirements for cyberspace operations both as supported and supporting commands in support of this campaign planning effort.

d. *Global Force Management Implementation Guidance* **(GFMIG)**

(1) The GFMIG provides SecDef's direction for global force management (GFM) to manage forces from a global perspective. It provides the specific direction for force assignment, apportionment, and allocation processes enabling SecDef to make risk-informed decisions regarding the distribution of US Armed Forces among the CCDRs.

(2) The GEF; GFMIG; and CJCSM 3130.06, *Global Force Management Allocation Policies and Procedures,* guide the GFM allocation process in support of CCMD force requirements. GFM processes align force apportionment, assignment, and allocation methodologies in support of the DSR and joint force availability requirements.

8. Joint Strategic Planning System

a. The JSPS is the primary system by which the CJCS carries out USC-assigned statutory responsibilities. The JSPS enables the CJCS to conduct assessments; provide military advice to the President, SecDef, NSC, and Homeland Security Council (HSC); and assist the President and SecDef in providing strategic direction to the US Armed Forces. The NMS and JSCP are core strategic guidance documents that provide CJCS direction and policy essential to the achievement of NSS objectives by augmenting the strategic direction provided in the UCP, GEF, and other Presidential directives. Other elements of JSPS, such as the CJCS risk assessment, the joint strategy review, and the annual joint assessment (AJA), inform decision making and identify new contingencies that may warrant planning and the commitment of resources. Figure II-1 illustrates these relationships.

The JSPS is described in detail in CJCSI 3100.01, Joint Strategic Planning System.

b. **Strategic Direction.** The President, SecDef, and CJCS use strategic direction to communicate their broad goals and issue-specific guidance to DOD. It provides the common thread that integrates and synchronizes the planning activities and operations of the JS, CCMDs, Services, joint forces, combat support agencies (CSAs), and other DOD agencies. It provides purpose and focus to the planning for employment of military force. Strategic direction identifies a desired military objective or end state; national-level planning assumptions; and national-level constraints, limitations, and restrictions. In addition to previously mentioned documents, additional strategic direction will emerge as orders or as part of the iterative plans dialogue reflected in APEX.

(1) **Policy and Strategic Assumptions.** Strategic guidance and specific strategic direction should include specific assumptions US leadership is willing to make for each planning effort. These assumptions should cover both domestic and international unknowns in order to better define the OE in which the commander is expected to operate. Similarly, the commander should identify and question strategic assumptions to determine if they are reasonable and offer suggestions for improvements and clarification.

(2) **Policy and Political Limitations.** The President and SecDef (or representatives) provide the commander and the command planning team any limitations (constraints or restraints) they expect will be imposed on the planning problem. These could be mandates for partner (or allied) participation, restrictions on military personnel levels, or expected basing limitations.

c. **NMS.** The NMS, derived from the NSS and DSR, prioritizes and focuses the efforts of the Armed Forces of the United States while conveying the CJCS's direction with regard to the OE and the necessary military actions to protect national security interests. The NMS defines the national military objectives (ends), how to accomplish these objectives (ways), and addresses the military capabilities required to execute the strategy (means). The NMS provides focus for military activities by defining a set of interrelated military objectives and joint operating concepts from which the Service Chiefs and CCDRs identify desired capabilities and against which the CJCS assesses risk.

Providing for the Direction of the Armed Forces

The National Military Strategy is the foundation for strategic integration; command and control; strategy and planning; programming and budgeting; and assessments.

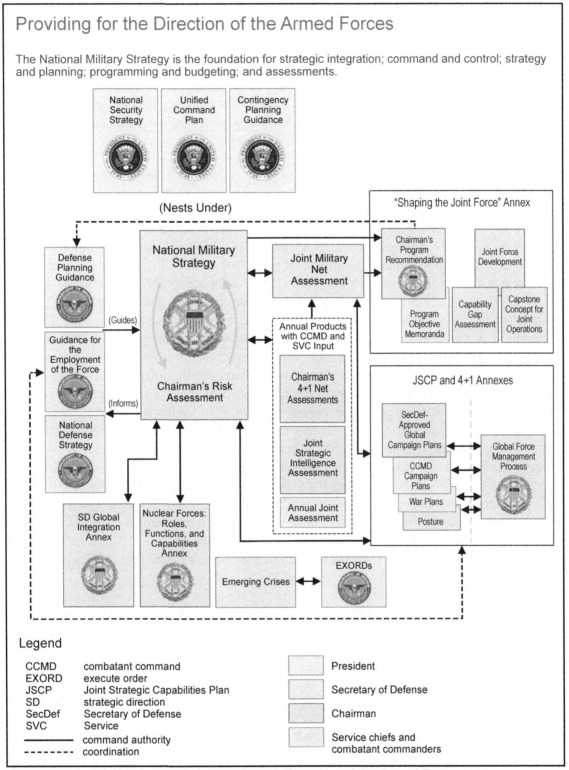

Figure II-1. Providing for the Direction of the Armed Forces

d. **JSCP.** The JSCP is the primary document in which the CJCS carries out his statutory responsibility for providing unified strategic direction to the Armed Forces. The JSCP provides military strategic and operational guidance to CCDRs, Service Chiefs,

CSAs, and applicable DOD agencies for preparation of plans based on current military capabilities. It implements the planning guidance provided in the GEF and the joint planning activities and products that accomplish that guidance. In addition to communicating to the CCMDs' specific planning guidance necessary for planning, the JSCP operationalizes the strategic vision described in the NMS and nests with the strategic direction delineated by the NSS, DSR, and the DOD's planning and resourcing guidance provided in the GEF. The JSCP also provides integrated planning guidance and direction for planners.

The JSCP is described in detail in CJCSI 3110.01, (U) 2015 Joint Strategic Capabilities Plan (JSCP).

e. **GFMIG.** The GFMIG documents force planning and execution guidance and show assignment of forces in support of the UCP. GFM aligns force assignment, apportionment, and allocation methodologies in support of the DSR and GEF, joint force availability requirements, and joint force assessments. It provides comprehensive insights into the global availability of US military resources and provides senior decision makers a process to quickly and accurately assess the impact and risk of proposed changes in force assignment, apportionment, and allocation. JS prepares the document for SecDef approval, with the Joint Staff J-8 [Director for Force Structure, Resource, and Assessment] overseeing the assignment and apportionment of forces and the Joint Staff J-3 [Operations Directorate] overseeing the allocation of forces. It is updated every two years and approved by SecDef. The GFMIG informs planners of the processes for distributing forces globally. It provides SecDef direction to the Secretaries of the Military Departments for assigning forces to CCDRs in order to accomplish their assigned missions, specifies the allocation process that provides access to forces and capabilities when assigned mission requirements exceed the capacity and/or capability of the assigned and currently allocated forces, includes apportionment guidance to facilitate planning, and informs the joint force structure and capability assessment processes. The assignment tables in the GFMIG and Forces for Unified Commands Memorandum serve as the record of force assignments. SecDef's decision to allocate forces is ordered in the *Global Force Management Allocation Plan* (GFMAP).

See Appendix E, "Global Force Management," for additional information and descriptions.

9. **Combatant Commanders**

a. **Planning Organization.** At the CCMD level, a joint planning group (JPG), operational planning group, or operational planning team (OPT) is typically established to direct planning efforts across the command, including implementation of plans and orders.

b. **Strategic Estimate.** The CCDR and staff, with input from subordinate commands and supporting commands and agencies, prepare a strategic estimate by analyzing and describing the political, military, economic, social, information, and infrastructure (PMESII) factors and trends, and the threats and opportunities that facilitate or hinder achievement of the objectives over the timeframe of the strategy.

(1) The strategic estimate is a tool available to commanders as they develop plans. CCDRs use strategic estimates developed in peacetime to facilitate the employment of military forces across the range of military operations. The strategic estimate is more comprehensive in scope than estimates of subordinate commanders, encompasses all aspects of the CCDR's OE, and is the basis for the development of the GCC's theater strategy.

(2) The CCDR, the CCDR's staff, and supporting commands and agencies evaluate the broad strategic-level factors that influence the theater strategy.

(3) The estimate should include an analysis of strategic direction received from the President, SecDef, or the authoritative body of a MNF; an analysis of all states, groups, or organizations in the OE that may threaten or challenge the CCMD's ability to advance and defend US interests in the region; visualization of the relevant geopolitical, geoeconomic, and cultural factors in the region; an evaluation of major strategic and operational challenges facing the CCMD; an analysis of known or anticipated opportunities the CCMD can leverage; and an assessment of risks inherent in the OE.

(4) The result of the strategic estimate is a visualization and better understanding of the OE to include allies, partners, neutrals, enemy combatants, and adversaries. The strategic estimate process is continuous and provides input used to develop strategies and implement plans. The broad strategic estimate is also the starting point for conducting the commander's estimate of the situation for a specific operation.

(5) Supported and supporting CCDRs and subordinate commanders all prepare strategic estimates based on assigned tasks. CCDRs who support multiple JFCs prepare estimates for each supporting operation.

See Appendix B, "Strategic Estimate," for a notional strategic estimate format.

c. **CCMD Strategies.** A strategy is a broad statement of the commander's long-term vision. It is the bridge between national strategic guidance and the joint planning required to achieve national and command objectives and attain end states. Specifically, it links CCMD activities, operations, and resources to USG policy and strategic guidance. A strategy should describe the ends as directed in strategic guidance and the ways and means to attain them. A strategy should begin with the strategic estimate. Although there is no prescribed format for a strategy, it may include the commander's vision, mission, challenges, trends, assumptions, objectives, and resources. CCDRs employ strategies to align and focus efforts and resources to mitigate and prepare for conflict and contingencies, and support and advance US interests. To support this, strategies normally emphasize security cooperation activities, force posture, and preparation for contingencies. Strategies typically employ military engagement, close cooperation with DOS, embassies, and other USG departments and agencies. A strategy should be informed by the means or resources available to support the attainment of designated end states and may include military resources, programs, policies, and available funding. CCDRs publish strategies to provide guidance to subordinates and supporting commands/agencies and improve coordination with other USG departments and agencies and regional partners. A CCDR operationalizes a strategy through a campaign plan (see Figure II-2).

For additional information on interagency considerations, see JP 3-08, Interorganizational Cooperation, *and USAID's* 3D Planning Guide: Diplomacy, Development, Defense.

10. Commander's Communication Synchronization

a. **Commander's communication synchronization** is the process to coordinate and synchronize narratives, themes, messages, images, operations, and actions to ensure their integrity and consistency to the lowest tactical level across all relevant communication activities.

b. Within the USG, DOS has primary responsibility for communication synchronization oversight. It is led by the Under Secretary for Public Diplomacy and Public Affairs and is the overall mechanism by which the USG coordinates public diplomacy across the interagency community. A key product of this committee is the US National Strategy for Public Diplomacy and Strategic Communication. This document

---

Additional Sources of Strategic Guidance

- National Security Strategy
- National Strategy for Combating Terrorism
- National Strategy for Public Diplomacy and Strategic Communication
- National Counterintelligence Strategy
- National Intelligence Strategy
- National Strategy to Combat Terrorist Travel
- National Strategy to Secure Cyberspace
- National Strategy for Homeland Security
- National Strategy for Maritime Security
- National Strategy for Information Sharing
- National Strategy for Pandemic Influenza
- National Strategy for Physical Protection of Critical Infrastructure
- National Strategy for Countering Biological Threats

List is not all inclusive.

Figure II-2. Additional Sources of Strategic Guidance

provides USG-level guidance, intent, strategic imperatives, and core messages under which DOD can nest its themes, messages, images, and activities.

c. The US military plays an important supporting role in communication synchronization, primarily through commander's communication synchronization, public affairs, and defense support to public diplomacy. Communication synchronization considerations should be included in all joint planning for military operations from routine, recurring military activities in peacetime through major operations.

d. Every JFC has the responsibility to coordinate, integrate, and synchronize **communications** to support planning and execution of a coherent national effort.

e. In addition to synchronizing the communication activities within the joint force, an effective communication synchronization effort is developed in concert with other USG departments and agencies, partner nations, and NGOs as appropriate. CCDRs should develop staff procedures for implementing communication synchronization guidance into all joint planning and targeting processes as well as collaborative processes for integrating communication synchronization activities with nonmilitary partners and subject matter experts.

See JP 1, Doctrine for the Armed Forces of the United States; *JP 3-0*, Joint Operations; *JP 3-61*, Public Affairs; *and Joint Doctrine Note 2-13*, Commander's Communication Synchronization, *for additional information.*

SECTION B. APPLICATION OF GUIDANCE

11. Joint Planning and Execution Community

a. The headquarters, commands, and agencies involved in joint planning or committed to a joint operation are collectively termed the **JPEC**. Although not a standing or regularly meeting entity, the JPEC consists of the stakeholders shown in Figure II-3.

(1) The **supported CCDR** has primary responsibility for all aspects of a task assigned by the GEF, the JSCP, or other joint planning directives. In the context of joint planning, the supported commander can initiate planning at any time based on command authority or in response to direction or orders from the President, SecDef, or CJCS. The designated supporting commanders provide planning assistance, forces, or other resources to a supported commander, as directed.

(2) **Supporting commanders** provide forces, assistance, or other resources to a supported commander in accordance with the principles set forth in JP 1, *Doctrine for the Armed Forces of the United States*. Supporting commanders prepare supporting plans as required. A commander may be a supporting commander for one operation while being a supported commander for another.

b. The President, with the advice and assistance of the NSC and CJCS, issues policy and strategic direction to guide the planning efforts of DOD and other USG departments and agencies that represent all of the instruments of national power. SecDef, with the

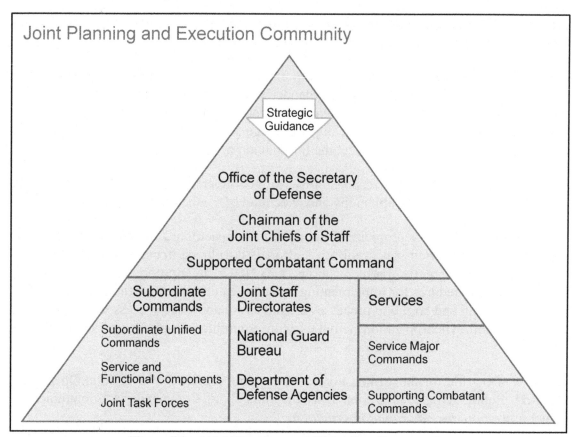

Figure II-3. Joint Planning and Execution Community

advice and assistance of the CJCS, organizes the JPEC for joint planning by establishing appropriate command relationships among the CCDRs and by establishing appropriate support relationships between the CCDRs and the CSAs for that portion of their missions involving support for operating forces. A supported commander is identified for specific planning tasks, and other JPEC stakeholders are designated as appropriate. This process provides for increased unity of command in the planning and execution of joint operations and facilitates unity of effort within the JPEC.

See CJCSI 3141.01, Management and Review of Joint Strategic Capabilities Plan (JSCP)-Tasked Plans, *for a more complete discussion of the JPEC. See JP 1,* Doctrine for the Armed Forces of the United States, *and JP 3-0,* Joint Operations, *for a more complete discussion of command relationships.*

12. Adaptive Planning and Execution Enterprise

a. APEX integrates the planning activities of the JPEC and facilitates the transition from planning to execution. The APEX enterprise operates in a networked, collaborative environment, which facilitates dialogue among senior leaders, concurrent and parallel plan development, and collaboration across multiple planning levels. Strategic direction and continuous dialogue between senior leaders and planners facilitate an early understanding of the situation, problems, and objectives. The intent is to develop plans that contain military options for the President and SecDef as they seek to shape the environment and respond to contingencies. This facilitates responsive plan development that provides up-to-date planning and plans for civilian leaders. The APEX enterprise also promotes involvement with other USG departments and agencies and multinational partners.

b. While joint planning has the inherent flexibility to adjust to changing requirements, APEX incorporates policies and procedures to facilitate a more responsive planning process. APEX fosters a shared understanding of the current OE and planning through frequent dialogue between civilian and military leaders to provide viable military options to the President and SecDef. Continuous assessment and collaborative technology provide increased opportunities for consultation and updated guidance during the planning and execution processes.

c. APEX encompasses four operational activities, four planning functions, seven execution functions, and a number of related products (see Figure II-4). Each of these planning functions will include IPRs as necessary throughout planning and execution. IPR participants are based on the requirements of the plan. For example, plans directed by the GEF or JSCP generally require SecDef-level review, while plans directed by a CCDR may require only CCDR-level review.

d. IPRs are an iterative dialogue among civilian and military leaders at the strategic level to gain a shared understanding of the situation, inform leadership, and influence planning. Topics such as planning assumptions, interagency and multinational participation guidance, supporting and supported activity requirements, desired objectives, key capability shortfalls, acceptable levels of risk, and SecDef decisions are typically discussed. Further, IPRs expedite planning by ensuring the plan addresses the most current strategic assessments and objectives.

See CJCS Guide 3130, Adaptive Planning and Execution Overview and Policy Framework, *for a more complete discussion of the APEX enterprise. CJCSI 3141.01,* Management and Review of Joint Strategic Capabilities Plan (JSCP)-Tasked Plans, *discusses IPRs in more detail.*

13. Operational Activities

a. Operational activities are comprised of a sustained cycle of situational awareness, planning, execution, and assessment activities that occur continuously to support leader decision-making cycles at all levels of command.

b. Situational Awareness

(1) Situational awareness addresses procedures for describing the OE, including threats to national security. This occurs during continuous monitoring of the national and international political and military situations so CCDRs, JFCs, and their staffs can determine and analyze emerging crises, notify decision makers, and determine the specific nature of the threat. Persistent or recurring theater military engagement activities contribute to maintaining situational awareness.

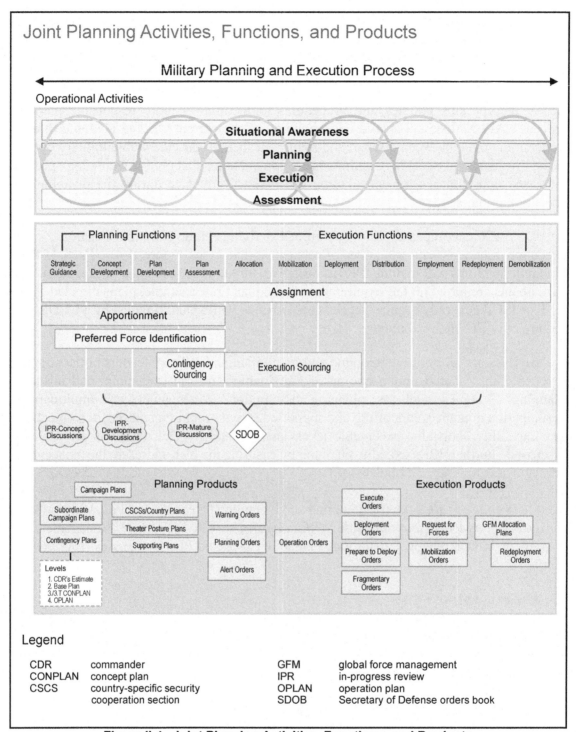

Figure II-4. Joint Planning Activities, Functions, and Products

(2) Situational awareness encompasses activities such as monitoring the global situation, identifying that an event has occurred, recognizing the event is a problem or a potential problem, reporting the event, and reviewing enduring and emerging warning concerns and the CCMD's running intelligence estimate (based on continuous joint intelligence preparation of the operational environment [JIPOE]). An event is a national or international occurrence assessed as unusual and viewed as potentially having an

adverse impact on US national interests and national security. The recognition of the event as a problem or potential problem follows from the observation.

c. **Planning**

(1) Planning translates strategic guidance and direction into campaign plans, contingency plans, and OPORDs. Joint planning may be based on defined tasks identified in the GEF and the JSCP. Alternatively, joint planning may be based on the need for a military response to an unforeseen current event, emergency, or time-sensitive crisis.

(2) Planning for contingencies is normally tasked in the JSCP based on the GEF or other directive. Planners derive assumptions needed to continue planning and reference the force apportionment tables to provide the number of forces reasonably expected to be available.

(3) Planning for crises is initiated to respond to an unforeseen current event, emergency, or time-sensitive crisis. It is based on planning guidance, typically communicated in orders (e.g., alert order [ALERTORD], warning order [WARNORD], PLANORD), and actual circumstances. Supported commanders evaluate the availability of assigned and currently allocated forces to respond to the event. They also determine what other force requirements are needed and begin putting together a rough order of magnitude force list.

d. **Execution**

(1) Execution begins when the President or SecDef authorizes the initiation of a military operation or other activity. An execute order (EXORD), or other authorizing directive, is issued by the CJCS at the direction of the President or SecDef to initiate or conduct the military operations. Depending upon time constraints, an EXORD may be the only order a CCDR or subordinate commander receives. The EXORD defines the time to initiate operations and conveys guidance not provided earlier.

(2) The CJCS monitors the deployment and employment of forces, makes recommendations to SecDef to resolve shortfalls, and tasks directed actions by SecDef and the President to support the successful execution of military operations. Execution continues until the operation is terminated or the mission is accomplished. In execution, based on continuous assessment activities, the planning process is repeated as circumstances and missions change.

(3) The supported CCDR monitors the deployment, distribution, and employment of forces; measures task performance and progress toward mission accomplishment; and adapts and adjusts operations as required to achieve the objectives and attain the end state. This continual assessment and adjustment of operations creates an organizational environment of learning and adaptation. This adaptation can range from minor operational adjustments to a radical change of approach. When fundamental changes have occurred that challenge existing understanding or indicate a shift in the OE/problem, commanders and staffs may develop a new operational approach that recognizes that the initial problem has changed, thus requiring a different approach to

solving the problem. The change to the OE could be so significant that it may require a review of the global strategic, theater strategic, and military end states and discussions with higher authority to determine whether the end states are still viable.

(4) Changes to the original plan may be necessary because of tactical, intelligence, or environmental considerations; force and non-unit cargo availability; availability of strategic lift assets; and port capabilities. Therefore, ongoing refinement and adjustment of deployment requirements and schedules and close coordination and monitoring of deployment activities are required.

(5) The CJCS-issued EXORD defines D-day [the unnamed day on which operations commence or are scheduled to commence] and H-hour [the specific time an operation begins] and directs execution of the OPORD. Date-time groups are expressed in universal time. While OPORD operations commence on the specified D-day and H-hour, deployments providing forces, equipment, and sustainment to support such are defined by C-day [an unnamed day on which a deployment operation begins], and L-hour [a specific hour on C-day at which a deployment operation commences or is to commence]. The CJCS's EXORD is a record communication that authorizes execution of the COA approved by the President or SecDef and detailed in the supported commander's OPORD. It may include further guidance, instructions, or amplifying orders. In a fast-developing crisis, the EXORD may be the first record communication generated by the CJCS. The record communication may be preceded by a voice authorization. The issuance of the EXORD is time sensitive. The format may differ depending on the amount of previous correspondence and the applicability of prior guidance. CJCSM 3122.01, *Joint Operation Planning and Execution System (JOPES) Volume I (Planning Policies and Procedures)*, contains the format for the EXORD. Information already communicated in previous orders should not be repeated unless previous orders were not made available to all concerned. The EXORD need only contain the authority to execute the operation and any additional essential guidance, such as D-day and H-hour.

(6) Throughout execution, JS, JFPs, Services, CCDRs, and CSAs monitor movements, assess accomplishment of tasks, and resolve shortfalls as necessary. This allows guidance to be changed and the plan to be modified, if necessary.

(7) The supported commander issues an OPORD to subordinate and supporting commanders prior to or upon receipt of an EXORD issued by the CJCS at the direction of the President or SecDef. It may provide detailed planning guidance resulting from updated or amplifying orders, instructions, or guidance that the EXORD does not cover. Supporting commanders may develop OPORDs in support of the supported commander's OPORD. The supported commander also implements an operation assessment, which evaluates the progress toward or achievement of military objectives. This assessment informs the commanders' recommendation to the President and SecDef of when to terminate a military operation. If significant changes in the OE or the problem are identified which call into question viability of the current operational approach or objectives, the supported commander should consult with subordinate and supporting commanders and higher authority.

(8) Following the GFM allocation process as detailed in CJCSM 3130.06, *(U) Global Force Management Allocation Policies and Procedures,* the supported CCDR's approved and validated force requests that have been allocated by the SecDef's decision are entered in the GFMAP annexes. The JFPs subsequently release GFMAP annex schedules reflecting specific deployment directions. The GFMAP Annexes (A and D) and Annex Schedules (B and C) serve as the deployment order (DEPORD) for specific FPs to allocate forces.

(9) GCCs coordinate with USTRANSCOM, other supporting CCDRs, JS, and FPs to provide an integrated transportation plan from origin to destination. The transportation component commands of USTRANSCOM (Army Military Surface Deployment and Distribution Command, and Navy Military Sealift Command) coordinate common-user land and sea movements while the Air Force Air Mobility Command coordinates common user air movements for the supported GCC's time-phased force and deployment data (TPFDD). The GCCs control the flow of requirements into and out of their theater, using the appropriate TPFDD validation process, in which both supporting and supported CCMDs' staff and Service components validate unit line numbers throughout the flow.

e. **Operation Assessment**

(1) Assessment determines the progress of the joint force toward mission accomplishment. Throughout the four planning functions, assessment involves comparing desired conditions of the OE with actual conditions to determine the overall effectiveness of the campaign or operation. More specifically, assessment helps the JFC measure task performance, determine progress toward or regression from accomplishing a task, creating an effect, achieving an objective, or attaining an end state; and issue the necessary guidance for change to guide forward momentum.

(2) Assessment is a continuous operation activity in both planning and execution functions and informs the commander's decision making. It determines whether current actions and conditions are creating the desired effects and changes in the OE towards the desired objectives. Before changes in the OE can be observed, a baseline or initial assessment is required. As follow-on assessments occur, historical trends can aid the analysis and provide more definitive and reliable measures and indicators of change. Assessment helps commands analyze changes in the OE, strategic guidance, or the challenges facing the joint force in order to adapt and update plans and orders to effectively achieve desired objectives.

(3) During planning, analysis associated with assessment helps facilitate greater understanding of the current conditions of the OE as well as identify how the command will determine the achievement of objectives if the plan is executed.

(4) During execution, assessment helps the command evaluate the progress or regression toward mission accomplishment, and then adapt and adjust operations as required to reach the desired end state (or strategic objectives). This analysis and adjustment of operations creates an organizational environment of learning and adaptation.

Adaptation can range from minor operational adjustments to a radical change of approach, including termination of the operation. When fundamental changes have occurred that challenge existing understanding or indicate a shift in the OE, commanders and staffs may develop a new operational approach that recognizes that the initial problem has changed, thus requiring a different approach toward the solution. The change to the OE could be so significant that it may require a review of the national strategic, theater strategic, and military objectives and discussions with higher authority to determine whether the military objectives or national strategic end states are still viable.

For more information on operation assessment, see Chapter VI, "Operation Assessment."

14. Planning Functions

a. The four planning functions of strategic guidance, concept development, plan development, and plan assessment are generally sequential, although often run simultaneously in order to deepen the dialogue between civilian and military leaders and accelerate the overall planning process. SecDef, CJCS, or the CCDR may direct the planning staff to refine or adapt a plan by reentering the planning process at any of the earlier functions. The time spent accomplishing each activity and function depends on the circumstances. In time-sensitive cases, planning functions may be compressed and decisions reached in an open forum. Orders may be combined and initially communicated orally.

b. **Strategic Guidance.** Strategic guidance initiates planning, provides the basis for mission analysis, and enables the JPEC to develop a shared understanding of the issues, OE, objectives, and responsibilities.

See Chapter IV, "Operational Art and Operational Design," for more details on the development of the commander's approach and operational concept.

(1) The CCDR provides input through sustained civilian-military dialogue that may include IPRs. The CCDR crafts objectives that support national strategic objectives with the guidance and consent of the SecDef; if required, the CJCS offers advice to SecDef. This process begins with an analysis of existing strategic guidance such as the GEF and JSCP or a CJCS WARNORD, PLANORD, or ALERTORD issued in a crisis. It includes mission analysis, threat assessment, and development of assumptions, which as a minimum, will be briefed to SecDef during the strategic guidance IPR.

(2) Some of the primary end products of the strategic guidance planning function are assumptions, ID of available/acceptable resources, conclusions about the strategic and OE (nature of the problem), strategic and military objectives, and the supported commander's mission.

(3) The CCDR will maintain dialogue with DOD leadership to ensure a common understanding of the above topics and alignment of planning to date. This step can be iterative, as the CCDR consults with the staff to identify concerns with or gaps in the guidance.

c. **Concept Development.** During planning, the commander develops several COAs, each containing an initial CONOPS that should identify major capabilities and authorities required and task organization, major operational tasks to be accomplished by components, a concept for employment and sustainment, and assessment of risk. Each COA may contain embedded multiple alternatives to accomplish designated objectives as conditions change (e.g., OE, problem, strategic direction). In time-sensitive situations, a WARNORD may not be issued, and a PLANORD, ALERTORD, or EXORD might be the first directive the supported commander receives with which to initiate planning. Using the guidance included in the directive and the CCDR's mission statement, planners solicit input from supporting and subordinate commands to develop COAs based upon the outputs of the strategic guidance planning function.

See Chapter V, "Joint Planning Process," for more details on COA development.

(1) During concept development, if an IPR is required, the commander outlines the COA(s) and makes a recommendation to higher authority for approval and further development.

(a) The commander recommends a COA that is most appropriate for the situation.

(b) Concept development should consider a range of COAs that integrate robust options to provide greater flexibility and to expedite transition during a crisis. CCDRs should be prepared to continue to develop multiple COAs to provide national-level leadership options should the crisis develop.

(c) For CCMD campaign plans, CCDRs should address resource requirements, expected changes in the environment, and how each COA supports achieving national objectives.

(d) The commander also requests SecDef's guidance on interorganizational planning and coordination and makes appropriate recommendations based on the interorganizational requirements identified during mission analysis and COA development.

(2) **One of the main** products from the concept development planning function is approval for continued development of one or more COAs. Detailed planning begins upon COA approval in the concept development function.

d. **Plan Development.** This function is used to develop a feasible plan or order that is ready to transition into execution. This function fully integrates mobilization, deployment, employment, sustainment, conflict termination, redeployment, and demobilization activities through all phases of the plan. When the CCDR believes the plan is sufficiently developed to become a plan of record, the CCDR briefs the final plan to SecDef (or a designated representative) for approval.

See Chapter V, "Joint Planning Process," for more details on completing the plan. See CJCSI 3141.01, Management and Review of Joint Strategic Capabilities Plan (JSCP)-Tasked Plans, *for more information on topics to be discussed during reviews at each stage.*

e. **Plan Assessment (Refine, Adapt, Terminate, Execute [RATE]).** Commanders continually review and evaluate the plan; determine one of four possible outcomes: refine, adapt, terminate, or execute; and then act accordingly. Commanders and the JPEC continue to evaluate the situation for any changes that would require changes in the plan. The CCDR will brief SecDef during routine plan update IPRs of modifications and updates to the plan based on the CCDR's assessment of the situation, changes in resources or guidance, and the plan's ability to achieve the objectives and attain the end states.

(1) **Refine.** During all planning efforts, plan refinement typically is an orderly process that follows plan development and is part of the assessment function. Refinement is facilitated by continuous operation assessment to confirm changing OE conditions related to the plan or potential contingency. In a crisis, continuous operation assessment accommodates the fluidity of the crisis and facilitates continuous refinement throughout plans or OPORD development. Planners frequently adjust the plan or order based on evolving commander's guidance, results of force planning, support planning, deployment planning, shortfall ID, adversary or MNF actions, changes to the OE, or changes to strategic guidance. Based on continuous operation assessment, refinement continues throughout execution, with changes typically transmitted in the form of fragmentary orders (FRAGORDs) rather than revised copies of the plan or order.

(2) **Adapt.** Planners adapt plans when major modifications are required, which may be driven by one or more changes in the following: strategic direction, OE, or the problem facing the JFC. Planners continually monitor the situation for changes that would necessitate adapting the plan, to include modifying the commander's operational approach and revising the CONOPS. When this occurs, commanders may need to recommence the IPR process.

(3) **Terminate.** Commanders may recommend termination of a plan when it is no longer relevant or the threat no longer exists. For GEF- or JSCP-tasked plans, SecDef, with advice from the CJCS, is the approving authority to terminate a planning requirement.

(4) **Execute.** See paragraph 13d, "Execution."

15. Planning Products

Joint planning encompasses the preparation of a number of planning and execution-related products. While the planning process is the same for CCMD campaign, contingency, or crisis planning, the output or products may differ.

a. **Products for CCMD Campaign and Contingency Planning.** Contingency and CCMD campaign planning encompasses the preparation of plans that occur in non-crisis situations with a timeline generally not driven by external events. It is used to develop plans for a broad range of activities based on requirements identified in the GEF, JSCP, or other planning directives. **CCMD campaign plans** are the centerpiece of DOD's planning construct. They provide the means to translate strategic guidance into CCMD strategies and subsequently into executable activities. **CCMD** campaign plans provide the vehicle for linking current operations to contingency plans.

(1) **Campaign Plans.** A campaign is a series of related military operations aimed at accomplishing strategic and operational objectives within a given time and space. Planning for a campaign is appropriate when the contemplated military operations exceed the scope of a single operation. Thus, campaigns are often the most extensive joint operations in terms of time and other resources. CCDRs document the full scope of their campaigns in the set of plans that includes the campaign plan and all of its subordinate and supporting plans.

(a) CCDRs plan and conduct campaigns and operations, while Service and functional components conduct subordinate campaigns, operations, activities, battles, and engagements, not independent campaigns. GCCs or FCCs can plan and conduct subordinate campaigns or operations in support of another CCMD's campaign. While intended primarily to guide the use of military power, discussions and decisions at the national strategic level provide guidance for employing the different instruments of national power and should be included in the campaign plan; as should the efforts of various interorganizational partners, to achieve national strategic objectives.

(b) Campaign plans implement a CCDR's strategy by comprehensively and coherently integrating all its activities (actual) and contingency (potential) operations. A CCDR's strategy and resultant campaign plan should be designed to achieve prioritized campaign objectives and serve as the integrating framework that informs and synchronizes all subordinate and supporting planning and operations. Campaign plans also help the CCDR in identifying resources required for achieving the objectives and tasks directed in the GEF and JSCP for input into budget and force allocation requests.

(c) Daily operations and activities should be designed to achieve national strategic objectives; to deter and prepare for crises identified in the GEF, JSCP, and other guidance documents; and to mitigate the potential impacts of a contingency. The campaign plan is the primary vehicle for organizing, integrating, and executing security cooperation activities.

(d) Under this construct, plans developed to respond to contingencies are best understood as branches to the overarching campaign plan (functional or theater). They address scenarios that put one or more US strategic end states in jeopardy and leave the US no recourse other than to address the problem through military actions (Figure II-5). Military actions can be in response to many scenarios, including armed aggression, regional instability, a humanitarian crisis, or a natural disaster. Contingency plans should provide a range of military options, to include flexible deterrent options (FDOs) or flexible response options (FROs), and should be coordinated with the total USG response.

(e) USTRANSCOM synchronizes global distribution operations primarily by guiding GCCs in the development of their TDPs that support the CCDR's campaign plan and other OPLANs.

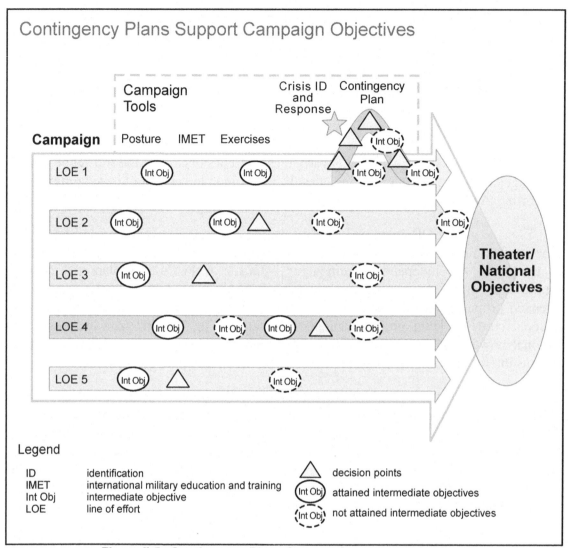

Figure II-5. Contingency Plans Support Campaign Objectives

(2) **Contingency Plans**

(a) Contingency plans are branches of campaign plans that are planned for potential threats, catastrophic events, and contingent missions without a crisis at-hand, pursuant to the strategic guidance in the UCP, GEF, and JSCP, and of the CCDR. A contingency is a situation that likely would involve military operations in response to natural and man-made disasters, terrorism, military operations by foreign powers, or other situations as directed by the President or SecDef.

(b) Planners develop plans from the best available information, using available forces and capabilities per the GFMIG, quarterly GFM apportionment tables, existing contracts, and task orders. Planning for contingencies is based on hypothetical situations and therefore relies heavily on assumptions regarding the circumstances that will exist when a crisis arises. Planning for a contingency encompasses the activities associated with the development of plans for the deployment, employment, sustainment, and redeployment of forces and resources in response to potential crises identified in joint strategic planning documents. An existing plan with a similar scenario may be used to initiate planning in an emergent crisis situation. To accomplish this, planners develop a CONOPS that details the assumptions; adversary forces; operation phases; prioritized missions; and force requirements, deployment, and positioning. Detailed, wargamed planning supports force requirements and training in preparation for the most likely operational requirements. It also enables rapid comparison of the hypothetical conditions, operation phases, missions, and force requirements of existing contingency plans to the actual requirements of an emergent crisis. Contingency planning allows the JPEC to develop understanding, as well as the analytical and planning expertise that can be useful during a crisis.

(c) If a situation develops outside of the strategic guidance development cycle of the GEF and JSCP that warrants a new plan that was not anticipated, the President or SecDef may issue direction through an SGS in response to the new situation. The CJCS implements the President's or SecDef's planning guidance into the appropriate orders or policy to direct the initiation of planning.

(d) Contingency plans are produced, reviewed, and updated periodically to ensure relevancy. This planning most often addresses contingencies where military options focus on combat operations. However, these plans also account for other types of joint military operations. In addition to plans addressing all phases, including those where military action may support other agencies, planning addresses contingencies where the military is in support from the onset. These include defense support of civil authorities, support to stabilization efforts, and foreign humanitarian assistance.

(e) There are four levels of planning detail for contingency plans, with an associated planning product for each level.

<u>1</u>. **Level 1 Planning Detail—Commander's Estimate.** This level of planning involves the least amount of detail and focuses on producing multiple COAs to address a contingency. The product for this level can be a COA briefing, command directive, commander's estimate, or a memorandum with a required force list. The commander's estimate provides SecDef with military COAs to meet a potential contingency. The estimate reflects the commander's analysis of the various COAs available to accomplish an assigned mission and contains a recommended COA.

<u>2</u>. **Level 2 Planning Detail—Base Plan (BPLAN).** A BPLAN describes the CONOPS, major forces, concepts of support, and anticipated timelines for completing the mission. It normally does not include annexes. A BPLAN may contain alternatives, including FDOs, to provide flexibility in addressing a contingency as it develops or to aid in developing the situation.

<u>3</u>. **Level 3 Planning Detail—CONPLAN.** A CONPLAN is an OPLAN in an abbreviated format that may require considerable expansion or alteration to convert it into a complete and detailed Level 4 OPLAN or an OPORD. It includes a plan summary, a BPLAN, and usually includes the following annexes: A (Task Organization), B (Intelligence), C (Operations), D (Logistics), J (Command Relations), K

(Communications), S (Special Technical Operations), V (Interagency Coordination), and Z (Distribution). If the development of a TPFDD is directed for the CONPLAN, the planning level is designated as 3T. A troop list and TPFDD would also require that an Annex E (Personnel) and Annex W (Operational Contract Support) be prepared.

For more information on OPLAN/CONPLAN format, see CJCSM 3130.03, Adaptive Planning and Execution (APEX) Planning Formats and Guidance, *and Appendix A, "Joint Operation Plan Format."*

4. **Level 4 Planning Detail—OPLAN.** An OPLAN is a complete and detailed plan containing a full description of the CONOPS, all applicable annexes to the plan including a time-phased force and deployment list (TPFDL), and a transportation-feasible notional TPFDD. The notional TPFDD phases unit requirements in the theater of operations at the times and places required to support the CONOPS. The OPLAN identifies the force requirements, functional support, and resources required to execute the plan and provide closure estimates for their flow into the theater. An OPLAN is normally prepared when:

a. The contingency is critical to national security and requires detailed prior planning.

b. The magnitude or timing of the contingency requires detailed planning.

c. Detailed planning is required to support multinational planning.

d. Detailed planning is necessary to determine force deployment, employment, sustainment, and redeployment requirements; determine available resources to fill identified requirements; and validate shortfalls.

(f) Contingency plans are created as part of a collaborative process with SecDef, OSD, CJCS, JCS, CCDRs, Services, and staffs of the entire JPEC for all contingencies identified in the GEF, JSCP, and other planning directives. Planning includes JPEC concurrent and collaborative joint planning activities. The JPEC reviews those plans tasked in the JSCP for SecDef approval. The USD(P) also reviews those plans for policy considerations in parallel with their review by the CJCS. CCDRs may request a JPEC review for any tasked or untasked plans that pertain to their AOR. CCDRs may direct the development of additional plans by their commands to accomplish assigned or implied missions.

(g) When directed by the President or SecDef through the CJCS, CCDRs convert level 1, 2, and 3 plans into level 4 OPLANs or into fully developed OPORDs for execution.

(3) **Cross-AOR Planning**

(a) When the scope of contemplated military operations exceeds the authority or capabilities of a single CCDR to plan and execute, the President, SecDef, or the CJCS, when designated by the President or SecDef, identify a CCDR to lead the planning for the designated strategic challenge or threat. The commander's assessment supporting this decision could be either the assessments of multiple CCDRs addressing a similar threat in their AORs or a single threat assessment from a CCDR addressing the threat from a global, cross-AOR, or functional perspective. Situations that may trigger this assessment range from combat operations that span UCP-designated boundaries to the threat of asymmetric attack that transits CCMD boundaries and functions and requires the strategic integration of the campaigns and operations of two or more CCDRs.

(b) Per Title 10, USC, SecDef may exercise responsibilities for overseeing the activities of the CCMDs through the CJCS. Such assignment by SecDef does not confer any command authority on the CJCS and does not alter CCDRs' responsibilities prescribed in Title 10, USC, Section 164(b)(2).

(c) **When designated,** the CJCS or delegated CCDR, with the authority of SecDef, issues a planning directive to the JPEC and may be tasked to lead the planning effort. The CJCS or delegated CCDR performs a mission analysis; issues initial global planning guidance based on national strategic objectives and priorities; and develops COAs in coordination with the affected CCMDs, Services, and CSAs. This COA mitigates operational gaps, seams, and vulnerabilities from a global perspective and develops an improved understanding of how actions in one AOR impact ongoing or potential plans and operations in other AORs. This will be achieved through a recommendation for the optimal allocation, prioritization, or reallocation of forces and capabilities required to develop a cohesive global CONOPS. These planning procedures will detail how CCDRs will employ forces and capabilities in support of another CCDR. The COA will be based largely on recommendations of the affected CCDRs. However, it should also assess the cumulative risk beyond a limited time horizon from a global perspective. These COAs may require refinement as initial planning apportionments are adjusted across the global CONOPS. Planners must be aware of competing requirements for potentially scarce strategic resources such as intelligence, surveillance, and reconnaissance (ISR) capabilities and transportation and ensure global planning is coordinated with GFM procedures.

(d) All planning should be collaborative and integrated. Integrated planning addresses complex threats that span multiple AORs and functional responsibilities and provides the President and SecDef a clear understanding of how the entire military, not just a portion, will respond to those threats. The CJCS or delegated CCDR is required to mitigate operational gaps, seams, and vulnerabilities and resolve the conflict over forces, resources, capabilities, or priorities from a global perspective. Employment of space, cyberspace, and special capabilities must be informed by risks, benefits, and tradeoff considerations. Early ID and submission of requests for forces and authorities and clear articulation of intent and risk can expedite decision making associated with employment of these capabilities.

(e) When directing the execution of a contingency plan or OPORD, the President or SecDef will also select a CCDR as the supported commander for implementation of the plan. The designated supported commander has primary responsibility for all aspects of a mission. In the context of planning, the supported commander leads integrated planning with supporting CCDRs to prepare plans or orders in response to higher headquarters requirements.

(4) **Supporting Plans.** Supporting CCDRs, subordinate JFCs, component commanders, and CSAs prepare supporting plans as tasked by the JSCP or other planning guidance. Commanders and staffs prepare supporting plans in CONPLAN/OPLAN format that follow the supported commander's concept and describe how the supporting commanders intend to achieve their assigned objectives and/or tasks. Supporting commanders and staffs develop these plans in collaboration with the supported commander's planners.

CJCSI 3141.01, Management and Review of Joint Strategic Capabilities Plan (JSCP)-Tasked Plans, *governs the formal review and approval process for campaign plans and level 1–4 plans.*

b. **Products of Planning in Crises**

(1) **Overview**

(a) A crisis is an incident or situation that typically develops rapidly and creates a condition of such diplomatic, economic, or military importance that the President or SecDef considers a commitment of US military forces and resources to achieve or defend national objectives. It may occur with little or no warning. It is fast-breaking and requires accelerated decision making. Sometimes a single crisis may spawn another crisis elsewhere, or there may be multiple crises occurring that concurrently impact two or more CCDRs. Furthermore, there may be a single threat with cross-AOR implications that simultaneously threaten two or more CCDRs. In this situation, supported and supporting command relationships may be fluid. Forces and capabilities committed to mitigate the emergent threat will require dynamic reallocation or reprioritization. These situations, which are increasingly the norm, further highlight the key role of integrated planning. While the planning and thought process are the same, planning in response to a crisis generally results in the publication of an order and the execution of an operation.

(b) Planning initiated in response to an emergent event or crisis uses the same construct as all other planning. However, steps may be compressed to enable the time-sensitive development of OPLANs or OPORDs for the deployment, employment, and sustainment of forces and capabilities in response to a situation that may result in actual military operations. While planning for contingencies is based on hypothetical situations and normally is conducted in anticipation of future events, planning in a crisis is based on circumstances that exist at the time planning occurs. When possible, planners should use previously prepared plans when the emergent crisis is similar. If unanticipated circumstances occur, and no previously developed plan proves adequate for the operational circumstances, then planning would begin from scratch. Regardless of whether a plan already exists, a similar plan will be modified, or planning for the emergent crisis will begin from scratch, for those crisis situations where the problem or threat affects more than one CCDR, the basic tenets of integrated planning would still apply. There are always situations arising in the present that might require a US military response. Such situations may approximate those previously planned for, although it is unlikely they would be identical, and sometimes they will be completely unanticipated. The time available to plan responses to such real-time events can be short. In as little as a few days, commanders and staffs may need to develop and approve a feasible COA with a notional TPFDD; publish the plan or order; prepare forces; make certain that scarce assets such as communications systems, lift, precision munitions, and ISR are sufficient; develop and execute an integrated intelligence plan [Annex B (Intelligence)]; and arrange sustainment for the employment of US military forces. Figure II-6 provides a comparison of planning for future contingencies and planning in a crisis.

(c) In a crisis, situational awareness is continuously fed by the latest all-source intelligence and operations reports as part of the continuous assessment of operational activities. An adequate and feasible military response in a crisis demands flexible procedures that consider time available, rapid and effective communications, and relevant previous planning products whenever possible.

(d) **In a crisis or time-sensitive situation, the CCDR reviews previously prepared plans for suitability.** The CCDR may refine or adapt these plans into an executable OPORD or develop an OPORD from scratch when no useful contingency plan exists.

(e) APEX planning functions, whether performed deliberately or in response to a crisis, use the same construct to facilitate unity of effort and the transition from planning to execution. These planning functions can be compressed or truncated in time sensitive conditions to enable the rapid exchange of information and analysis, the timely preparation of military COAs for consideration by the President or SecDef, and the prompt transmission of their decisions to the JPEC. Planning activities may be performed sequentially or concurrently, with supporting and subordinate plans or OPORDs being developed concurrently. The exact flow of activities is largely determined by the time available to complete the planning and by the significance of the crisis. The following paragraphs summarize the activities and interaction that occur in a compressed planning process such as a crisis. Refer to the CJCSM 3130 and 3122 series of publications, which address planning policies and procedures, for detailed procedures.

1. When the President, SecDef, or CJCS decides to develop military options, the CJCS issues a planning directive to the JPEC initiating the development of COAs and requesting that the supported commander submit a commander's estimate of the situation with a recommended COA to resolve the situation. Normally, the directive will be a WARNORD, but a PLANORD or ALERTORD may be used if the nature and timing of the crisis warrant accelerated planning. In a quickly evolving crisis, the initial WARNORD may be communicated verbally with a follow-on record copy to ensure the JPEC is kept informed. If the directive contains a force deployment preparation order or DEPORD, SecDef approval is required.

2. The amount of detail included in the WARNORD depends on the known facts and time available when issued. The WARNORD should describe the situation, establish command relationships, and identify the mission and any planning constraints. It may identify forces and strategic mobility resources, or it may request that

Contingency and Crisis Comparison

	Planning for a Contingency	Planning in a Crisis
Time available	As defined in authoritative directives (normally 6+ months)	Situation dependent (hours, days, up to 12 months)
Environment	Distributed, collaborative planning	Distributed, collaborative planning and execution
Facts and assumptions	Significant use of assumptions	Rely on facts and minimal use of assumption
JPEC involvement	Full JPEC participation (Note: JPEC participation may be limited for security reasons.)	Full JPEC participation (Note: JPEC participation may be limited for security reasons.)
APEX operational activities	Situational awareness Planning Assessment	Situational awareness Planning Execution Assessment
APEX functions	Strategic guidance Concept development Plan development Plan assessment	Strategic guidance Concept development Plan development Plan assessment
Document assigning planning task	CJCS issues: 1. JSCP 2. Planning directive 3. WARNORD (for short suspense planning)	CJCS issues: 1. WARNORD 2. PLANORD 3. SecDef-approved ALERTORD
Forces for planning	Apportioned in JSCP	Allocated in WARNORD, PLANORD, or ALERTORD.
Planning guidance	CJCS issues JSCP or WARNORD. CCDR issues PLANDIR and TPFDD LOI.	CJCS issues WARNORD, PLANORD, or ALERTORD. CCDR issues WARNORD, PLANORD, or ALERTORD and TPFDD LOI to subordinates, supporting commands, and supporting agencies.
COA selection	CCDR prepares COAs and submits to CJCS and SecDef for review. Specific COA may or may not be selected.	CCDR develops commander's estimate with recommended COA.
CONOPS approval	SecDef approves planning or directs additional planning or changes.	President/SecDef approve COA, disapproves or approves further planning.
Final planning product	Campaign plan. Level 1–4 contingency plan.	OPORD
Final planning product approval	CCDR submits final plan to CJCS for review and SecDef for approval.	CCDR submits final plan to President/SecDef for approval.
Execution document	Not applicable.	CJCS issues SecDef-approved EXORD. CCDR issues EXORD.
Output	Plan	Execution

Legend

ALERTORD	alert order	JSCP	Joint Strategic Campaign Plan
APEX	Adaptive Planning and Execution	LOI	letter of instruction
CCDR	combatant commander	OPORD	operations order
CJCS	Chairman of the Joint Chiefs of Staff	PLANDIR	planning directive
COA	course of action	PLANORD	planning order
CONOPS	concept of operations	SecDef	Secretary of Defense
EXORD	execute order	TPFDD	time-phased force and deployment data
JPEC	joint planning and execution community	WARNORD	warning order community

Figure II-6. Contingency and Crisis Comparison

the supported commander develop these factors. It may establish tentative dates and times to commence mobilization, deployment, or employment, or it may solicit the recommendations of the supported commander regarding these dates and times. The WARNORD should also identify any planning assumptions, restraints, or constraints the President or SecDef have identified to shape the response. If the President, SecDef, or CJCS directs development of a specific option or especially a COA, the WARNORD will describe the COA and request the supported commander's assessment. A WARNORD sample is in the CJCSM 3130.03, *Adaptive Planning and Execution (APEX) Planning Formats and Guidance.*

3. In response to the WARNORD, the supported commander, in collaboration with subordinate and supporting commanders and the rest of the JPEC, reviews existing joint contingency plans for applicability and develops, analyzes, and compares COAs and prepares a commander's estimate that provides recommendations and advice to the President, SecDef, or higher headquarters for COA selection. Based on the supported commander's guidance, supporting commanders begin their planning activities.

4. Although an existing plan almost never completely aligns with an emerging crisis, it can be used to facilitate rapid COA development and be modified to fit the specific situation. TPFDDs developed for specific plans are stored in the Joint Operation Planning and Execution System (JOPES) database and are made available to the JPEC for review.

5. The CJCS, in consultation with other members of the JCS and JPEC, reviews and evaluates the supported CCDR's estimate and provides recommendations and advice to the President and SecDef for COA selection. The supported CCDR's COAs may be accepted, refined, or revised, or a new COA(s) may have to be developed. The President or SecDef selects a COA and directs that detailed planning be initiated.

6. Upon receiving directions from the President or SecDef, the CJCS issues a SecDef-approved ALERTORD to the JPEC. The order is a record communication stating the President or SecDef has approved the detailed development of a military plan to help resolve the crisis. The contents of an ALERTORD may vary depending upon the crisis and amount of prior planning accomplished, but it should always describe the selected COA in sufficient detail to allow the supported commander, in collaboration with other members of the JPEC, to conduct the detailed planning required to deploy, employ, and sustain forces. However, the ALERTORD does not authorize execution of the approved COA.

7. The supported commander then develops an OPORD using the approved COA. The speed with which the OPORD is developed depends upon the amount of prior planning and the planning time available. The supported commander and subordinate commanders identify force requirements, contracted support requirements and management, existing contracts and task orders, and mobility resources, and describe the CONOPS in OPORD format. The supported commander reviews available assigned and allocated forces that can be used to respond to the situation and if a gap exists, submits a request for forces (RFF) to the JS for forces to be allocated. For a detailed description of

the GFM allocation process refer to CJCSM 3130.06, *(U) Global Force Management Allocation Policies and Procedures.*

8. The supported CCDR submits the completed OPORD for approval to SecDef or the President via the CJCS. The President or SecDef may decide to begin deployment in anticipation of executing the operation or as a show of resolve, execute the operation, place planning on hold, or cancel planning pending resolution by some other means. Detailed planning may transition to execution as directed or become realigned with continuous situational awareness, which may prompt planning product adjustments and/or updates.

9. Plan development continues after the President or SecDef's execution decision. When the crisis does not lead to execution, the CJCS provides guidance regarding continued planning.

(f) **Abbreviated Procedures.** The preceding discussion describes the activities sequentially. During a crisis, they may be conducted concurrently or compressed, depending on prevailing conditions. It is also possible that the President or SecDef may decide to commit forces shortly after an event occurs, thereby significantly compressing planning activities. Although the allocation process has standard timelines, they may be accelerated. No specific length of time can be associated with any particular planning activity. **Severe time constraints may require crisis participants to pass information verbally, including the decision to commit forces.** Verbal orders are followed up, as soon as practical, with written orders.

(2) **Joint Orders.** Upon approval, CCDRs and Services issue orders directing action (see Figure II-7). Formats for orders can be found in CJCSM 3130.03, *Adaptive Planning and Execution (APEX) Planning Formats and Guidance.* By the CJCS's direction, the JS J-3 develops, coordinates, and prepares APEX orders. Subsequently, the JS J-3 is responsible for preparing and coordinating the Secretary of Defense Orders Book to present recommendations to SecDef for decision.

(a) **WARNORD.** A WARNORD, issued by the CJCS and/or commander, is a planning directive that initiates the development and evaluation of military COAs by a supported commander and requests that the supported commander submit a commander's estimate. If the order contains the deployment of forces, SecDef's authorization is required.

(b) **PLANORD.** A PLANORD is a planning directive that provides essential planning guidance and directs the initiation of plan development before the directing authority approves a military COA.

(c) **ALERTORD.** An ALERTORD is a planning directive that provides essential planning guidance and directs the initiation of plan development after the directing authority approves a military COA. An ALERTORD does not authorize execution of the approved COA.

Joint Orders

	Order Type	Intended Action	Secretary of Defense Approval Required
Warning order	WARNORD	Initiates development and evaluation of COAs by supported commander. Requests commander's estimate be submitted.	No. Required when WARNORD includes deployment or deployment preparation actions.
Planning order	PLANORD	Begins planning for anticipated President or SecDef-selected COA. Directs preparation of OPORDs or contingency plan.	No. Conveys anticipated COA selection by the President or SecDef.
Alert order	ALERTORD	Begins execution planning on President or SecDef-selected COA. Directs preparation of OPORD or contingency plan.	Yes. Conveys COA selection by the President or SecDef.
Operation order	OPORD	Effect coordinated execution of an operation.	Specific to the OPORD.
Prepare to deploy order	PTDO	Increase/decrease deployability posture of units.	Yes (if allocates force). Refers to five levels of deployability posture.
Deployment/ redeployment order	DEPORD	Deploy/redeploy forces. Establish C-day/L-hour. Increase deployability. Establish joint task force.	Yes (if allocates force). Required for movement of unit personnel and equipment into combatant commander's AOR.
Execute order	EXORD	Implement President or SecDef decision directing execution of a COA or OPORD.	Yes.
Fragmentary order	FRAGORD	Issued as needed after an OPORD to change or modify the OPORD execution.	No.

Legend

AOR   area of responsibility
C-day   unnamed day on which a deployment operation begins
COA   course of actions
L-hour   specific hour on C-day at which deployment operation commences or is to commence
SecDef   Secretary of Defense

Figure II-7. Joint Orders

(d) **Prepare to Deploy Order (PTDO).** PTDOs are approved by SecDef for allocated forces and contained in the GFMAP. The supported CCDR may order their assigned forces to deploy or order them to be prepared to deploy via a DEPORD. A PTDO is an order to prepare a unit to increase the deployability posture of units on a specified timeline.

(e) **DEPORD.** A planning directive from SecDef, issued by the CJCS, authorizes the transfer and allocation of all forces among CCMDs, Services, and DOD agencies and specifies the authorities the gaining CCDR will exercise over specified forces to be transferred. The GFMAP is a global DEPORD for all allocated forces. FPs deploy or prepare forces to deploy on a specified timeframe as directed in the GFMAP. CJCSM

3130.06, *(U) Global Force Management Allocation Policies and Procedures,* and GFMIG discuss the DEPORD in more detail.

(f) **EXORD.** An EXORD is a directive to implement an approved military CONOPS. Only the President and SecDef have the authority to approve and direct the initiation of military operations. The CJCS, by the authority of and at the direction of the President or SecDef, may subsequently issue an EXORD to initiate military operations. Supported and supporting commanders and subordinate JFCs use an EXORD to implement the approved CONOPS.

(g) **OPORD.** An OPORD is a directive issued by a commander to subordinate commanders for the purpose of effecting the coordinated execution of an operation. Joint OPORDs are prepared under joint procedures in prescribed formats during a crisis.

(h) **FRAGORD.** A FRAGORD is a modification to any previously issued order. It is issued as needed to change an existing order or to execute a branch or sequel of an existing OPORD. It provides brief and specific directions that address only those parts of the original order that have changed.

CHAPTER III
STRATEGY AND CAMPAIGN DEVELOPMENT

> *"For to win one hundred victories in one hundred battles is not the acme of skill. To subdue the enemy without fighting is the acme of skill."*
>
> **Sun Tzu, *The Art of War***

1. Overview

DOD is tasked to conduct operations on a daily basis to aid in achieving national objectives. In turn, CCDRs are tasked to develop strategies and campaigns to shape the OE in a manner that supports those strategic objectives. They conduct their campaigns primarily through military engagement, operations, posture, and other activities that seek to achieve US national objectives and prevent the need to resort to armed conflict while setting conditions to transition to contingency operations when required. The CCMD strategies and campaign plans are nested within the framework of the NSS, DSR, and NMS and are conducted in conjunction with the other instruments of national power. Specific guidance to the commanders is found in the UCP, GEF, and JSCP. Strategy prioritizes resources and actions to achieve future desired conditions. It acknowledges the current conditions as its start point, but must look past the current conditions and envision a future, then plot the road to get there. Plans address detailed execution to implement the strategy. National strategy prioritizes the CCMD's efforts within and across theaters, functional, and global responsibilities; and considers all means and capabilities available in the CCMD's operations, activities, and investments to achieve the national objectives and complement related USG efforts over a specified timeframe (currently five years). In this construct, the CCDRs and their planners develop strategy and plan campaigns to integrate joint operations with national-level resource planning and policy formulation and in conjunction with other USG departments and agencies.

a. Description

(1) **Vision.** The CCDR develops a long-range vision that is consistent with the national strategy and US policy and policy objectives. The vision is usually not constrained by time or resources, but is bounded by the national policy.

(2) **Strategy.** Strategy is a broad statement of the CCDR's long-term vision guided by and prepared in the context of SecDef's priorities and within projected resources. Strategy links national strategic guidance to joint planning.

(a) The CCDR's strategy prioritizes the ends, ways, and means within the limitations established by the budget, GFM processes, and strategic guidance/direction. The strategy must address risk and highlight where and what level risk will be accepted and where it will not be accepted. The strategy's objectives are directly linked to the achievement of national objectives.

(b) Strategy includes a description of the factors and trends in the OE key to achieving the CCMD's objectives, the CCDR's approach to applying military power in concert with the other instruments of national power in pursuit of the objectives and the risks inherent in implementation.

(c) Strategy must be flexible to respond to changes in the OE, policy, and resources. Commanders and their staff assess the OE, as well as available ways, means, and risk then update the strategy as needed. It also recognizes when ends need updating either because the original ones have been attained or they are no longer applicable.

b. **Purpose of the CCDRs' Campaign Plans**

(1) The CCDRs' campaigns operationalize the CCDRs' strategies by organizing and aligning operations, activities, and investments with resources to achieve the CCDRs' objectives and complement related USG efforts in the theaters or functional areas.

(2) CCDRs translate the strategy into executable actions to accomplish identifiable and measurable progress toward achieving the CCDRs' objectives, and thus the national objectives. The achievement of these objectives is reportable to DOD leadership through IPRs and operation assessments (such as the CCDRs' annual input to the AJA).

(3) CCMD campaign plans integrate posture, resources, requirements, subordinate campaigns, operations, activities, and investments that prepare for, deter, or mitigate identified contingencies into a unified plan of action.

(4) The purpose of CCMD campaigns is to shape the OE, deter aggressors, mitigate the effects of a contingency, and/or execute combat operations in support of the overarching national strategy.

(a) Shaping the OE is changing the current conditions within the OE to conditions more favorable to US interests. It can entail both combat and noncombat operations and activities to establish conditions that support future US activities or operations, or validate planning assumptions.

(b) Deterrence activities, as part of a CCMD campaign, are those **actions** or operations executed specifically to alter adversaries' decision calculus. These actions or operations may demonstrate US commitment to a region, ally, partner, or principle. They may also demonstrate a US capability to deny an adversary the benefit of an undesired action. Theater posture and certain exercises are examples of possible deterrent elements of a campaign. These actions are the most closely tied elements of the campaign to contingency plans directed in the GEF and JSCP. Additional deterrence activities are associated with early phases of a contingency plan, usually directed and executed in response to changes in threat posture.

(c) A campaign can also set conditions that mitigate the impact of a possible contingency (see Figure III-1). Activities conducted as part of the campaign, such as posture and security cooperation (e.g., military engagement with allies and partners or

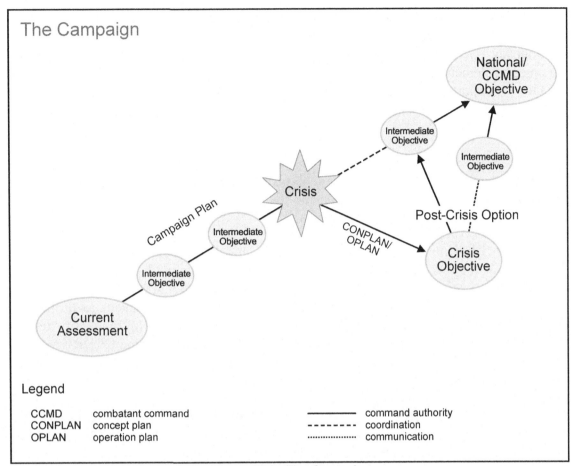

Figure III-1. The Campaign

building partner capacity and capability) can set the stage for more rapid, successful execution of a contingency plan if conflict erupts. Campaign activities can also validate planning assumptions used in the contingencies.

(d) A campaign can support stabilization. Where US national security objectives depend upon maintaining or reestablishing stability, stabilization links the application of joint force combat power and security assistance capabilities with the achievement of strategic and policy objectives. Stabilization efforts focus on the root causes of instability and mitigating the drivers of conflict for an affected host nation (HN), thus helping the HN reach a sustainable political settlement that allows societal conflicts to be resolved peacefully.

c. **Differences Between CCMD Campaign Plans and Contingency Plans**

(1) **CCMD campaigns plans** seek to shape the OE and achieve national objectives. They establish operations, activities, and investments the command undertakes to achieve specific objectives (set conditions) in support of national policy and objectives.

(a) CCMD campaigns are proactive and rarely feature a single measure of military success implying victory in the traditional sense.

(b) The campaign may include operations across the spectrum of conflict, to include ongoing combat operations, such as counterterrorism operations. In the event a contingency operation is executed, that operation is subsumed into the campaign and becomes an element the CCDR considers when identifying the impact of US operations on the OE, the opportunities to favorably affect the OE to achieve national-level and theater-level objectives, and examining MOEs that may impact the campaign's intermediate objectives.

(c) Campaign objectives must be continuously assessed to see if they are achieving the desired conditions. As objectives are achieved (or determined to be infeasible), the CCDR and planners update the campaign plan with new objectives and develop associated assessment measures.

(d) Unlike contingency plans, GEF- and JSCP-directed CCMD campaigns do not end with the achievement of military objectives and do not have end states to help determine termination criteria. Campaign plan objectives neither affirm nor imply military victories, but instead focus CCMD operations, activities, and investments to further US national security by supporting US national security objectives. It helps to identify desired OE conditions in order to focus campaign planning (the purpose of the CCDR's vision), with the understanding that campaign objectives and US interests may change as the OE evolves and policies change.

(e) The increasing global influence of non-state entities and hostile non-state actors challenges the process of identifying adversary and other actors' traditional centers of gravity (COGs) and vulnerabilities. Therefore, a campaign plan identifies mostly nonlethal means to favorably influence the OE to achieve specific intermediate objectives.

(f) Campaign plans seek to capitalize on the cumulative effect of multiple coordinated and synchronized operations, activities, and investments that cannot be accomplished by a single major operation.

(2) **Contingency plans** identify how the command might respond in the event of a crisis or the inability to achieve objectives. Contingency plans specifically seek to favorably resolve a crisis that either was not or could not be deterred or avoided by directing operations toward achieving specified objectives.

(a) Contingency plans have specified end states that seek to re-establish conditions favorable to the US. They react to conditions beyond the scope of the CCMD campaign plan.

(b) Contingency plans have an identified military objective and set of termination criteria. Upon terminating a contingency plan, military operations return to campaign plan execution. However, the post-contingency OE may require different or additional military activities to sustain new security conditions.

(c) Although campaign plan operations, activities, and investments can have deterrent effects, contingency plan deter activities specifically refer to actions for which separate and unique resourcing and planning are required. These actions are executed on order of the President or SecDef and generally entail specific orders for their execution and require additional resources allocated through GFM processes.

2. Campaign Planning

a. Campaigns and campaign planning follow the principles of joint operations while synchronizing efforts throughout the OE with all participants. Examples include:

(1) **Objective.** Clear campaign objectives must be articulated and understood across the joint force. Objectives are specified to direct every military operation toward a clearly defined, decisive, and achievable goal. Objectives may change as national and military leaders gain a better understanding of the situation, or they may occur because the situation itself changes. The JFC should remain sensitive to shifts in political goals necessitating changes in the military objectives toward attainment of the strategic end states.

(2) **Unity of Command.** Unity of command means all forces operate under a single commander with the requisite authority to direct all forces employed in pursuit of a common purpose. During multinational operations and interagency coordination, unity of command may not always be possible, but unity of effort, the coordination and cooperation toward common objectives, becomes paramount for successful unified action.

(3) **Economy of Force.** Economy of force is the judicious employment and distribution of forces to achieve campaign objectives.

(4) **Legitimacy.** Legitimacy maintains legal and moral authority in the conduct of operations. Legitimacy is based on the actual and perceived legality, morality, and rightness of actions from the perspectives of interested audiences.

See JP 3-0, Joint Operations, *for more information on the principles of joint operations.*

b. Campaign plans are informed by operation assessments that continuously measure progress or regression regarding clearly defined, measurable, and attainable intermediate objectives nested under campaign objectives. During the planning functions, planners can use a combination of operational design and JPP that asks four questions:

(1) What are the current conditions of the OE (where are you)?

(2) What are the future conditions you want to establish (where do you want to go; what is the objective)?

(3) How will you get there (resources and authorities)?

(4) How will you know that you have been successful (assessment)? (Assessment is not just measuring achievement of an intermediate or campaign objective. It also requires measuring the performance and the effects of joint activities to determine whether they can or will generate the desired effects or establish the desired conditions.)

See paragraph 6, "Assessing Theater and Functional Campaign Plans," and Appendix D, "Operation Assessment Plan (Examples)," for additional information on assessments.

c. Campaigns are informed by strategic guidance and the requirement to be ready to execute contingency plans. Throughout the four planning functions, beginning with mission analysis within JPP, the CCDR and staff develop and update the commander's critical information requirements (CCIRs). This concurrently complements assessment activities by including information requirements critical to addressing key assessment indicators, required contingency preparations, deterrent opportunities, and the critical vulnerabilities of all actors within the OE. Through backward planning, CCMDs identify precursor actions, campaign activities, and necessary authorities that should be executed (or provided) as part of the campaign to deter, prepare for, or mitigate contingencies outside of crisis conditions. If successfully conducted, the campaign mitigates the risk for conflict in the context of the directed contingency plan, sets conditions for more rapid and successful transition of the contingency plan to execution if conflict proves unavoidable, and sets conditions to forestall future crises.

d. The same construct of APEX operational activities and planning functions, processes, procedures, and tools is used by planners to develop contingency plans and campaign plans. The applications of these can be tailored.

(1) Because there is no military end state or termination criteria for a GEF- or JSCP-directed campaign, the objectives established in the plan are guideposts rather than goalposts and map a route in support of US objectives. The GEF- and JSCP-directed campaign plans do not seek to defeat an enemy in combat but to improve the OE in support of US national interests. As one objective is achieved, another should be designated.

(2) The frame of reference for the campaign plan must be critically examined. When trying to map a complex system, planners tend to map it from their point of view. The relationships and logic chains developed during planning will reflect their perspective. Other participants in the system, to include allies, partners, and adversaries, often come from different backgrounds with different rules and relationships, so the effects of US actions may not result in the desired conditions. What may seem like cooperation from a US perspective may appear to be coercion from the partner's perspective.

(3) Rather than having an enemy COG, the CCMD campaign plan may identify several COGs or areas the command may affect to achieve its objectives. Since the campaign addresses a large, complex problem, it may not be a single issue, but a confluence of several issues interacting that affect the OE. See Chapter IV, "Operational Art and Operational Design," for more information.

(4) **Lines of Effort (LOEs)**

(a) In GEF- and JSCP-directed campaigns, it is often easier to organize the campaign along LOEs. A LOE links multiple tasks and missions using the logic of purpose—cause and effect—to focus efforts toward establishing operational and strategic conditions. LOEs link intermediate objectives on a path to the military and hence the campaign objective. LOEs are used to visualize the relationships between conditions, campaign objectives and, by inference, the theoretical end state. Because a campaign is conditions-based and must be adaptive to events, LOEs indicate a route rather than a precise timetable of events. They indicate how, and in what order (and with what dependencies), it is envisaged that the activities of the joint force will contribute to the achievement of desired objectives.

(b) LOEs may intersect and interact. The campaign should identify how success or failure along a LOE will impact the lines of operation (LOOs) and other LOEs and, if necessary, how resources can be redirected to respond to unexpected effects (successes, failures, or unintended consequences) of operations on both its own and other LOEs.

(c) Everyone involved with conducting a campaign should know the intermediate and national objectives for the theater. Each tactical activity should be related to its military and theater objective through the LOE or LOO on which it is located. The operator or executor of each campaign activity should know both the success criteria of the specific task assigned as well as how that task relates to and supports the larger command objective.

For detailed discussion of LOOs and LOEs, see Chapter IV, "Operational Art and Operational Design," subparagraph 15g, "LOO and LOE."

e. Campaign plans will have some similarities with contingency plans.

(1) **Measurable and Time-Bound.** Campaign plans, like contingencies, must have measurable objectives and a process for associating CCMD actions to the changes in the OE. The commander must be able to identify within a directed time-span the ability to effect change and whether or not given actions successfully affected an associated change.

(2) **Changeable and Flexible**

(a) Campaigns must adapt to changes in the OE, other actors actions, and changes in resourcing and priorities based on national and defense priorities.

(b) However, a campaign should not necessarily change every time a commander or staff changes. Well-designed campaigns can withstand changes in foreseeable national leadership fluctuations in the US and by the countries addressed in the campaign. Continuity does not imply that changes in the COA or approach should be avoided; not adapting to the changes in the OE will lead to failure.

f. When a campaign addresses a persistent threat that spans multiple commands, such as terrorism, threats to space and cyberspace assets or capabilities, or distribution operations, the President or SecDef may designate coordinating authority to one CCDR to lead the planning effort, with execution accomplished across multiple CCMDs. CCMDs may identify those activities that support the overall plan through the development of a separate subordinate campaign plan or through inclusion in their overall campaign plan.

(1) The CCDR with coordinating authority coordinates planning efforts of CCDRs, Services, and applicable DOD agencies in support of the designated DOD global campaign plan. The phrase "coordinated planning" pertains specifically to planning efforts only and does not, by itself, convey authority to execute operations or direct execution of operations. Unless directed by SecDef, the CCDR responsible for leading the planning effort is responsible for aligning and harmonizing the CCMD campaign plans. Execution of the individual plans remains the responsibility of the GCC or FCC in whose UCP authority it falls.

(2) CCDRs develop subordinate campaign plans to satisfy the planning requirements of DOD global campaign plans. While these plans are designated subordinate plans, this designation does not alter current command relationships. GCCs remain the supported commanders for the execution of their plans unless otherwise directed by SecDef.

(3) If directed to develop or synchronize a DOD-wide campaign plan, the lead CCMD:

(a) Provides a common plan structure and strategic framework to guide and inform development of CCDR subordinate campaign plans, Services, CSAs, the National Guard Bureau, and other DOD agencies supporting plans and mitigate seams and vulnerabilities from a global perspective.

(b) Establishes a common process for the development of subordinate and supporting plans.

(c) Organizes and executes coordination and collaboration conferences in support of the global campaign to enhance development of subordinate and supporting plans consistent with the established strategic framework and to coordinate and conduct synchronization activities.

(d) Disseminates lessons learned to CCDRs, Services, and applicable DOD agencies. This includes the consolidation and standardization of planning efforts, products, and collaborative tools.

(e) Reviews and coordinates all subordinate and supporting plans to align them with the DOD global campaign plan.

(f) Assesses and provides recommendations to senior military and civilian leadership on the allocation of forces to coordinate the supported and supporting plans from a global perspective.

(g) Assesses supported and supporting plans and presents integrated force and capability shortfalls with potential sourcing options. These shortfalls and options inform SecDef of the challenges to executing the plan and the decisions that will likely be required should the plan transition to execution.

(h) Provides advice and recommendations to CCDRs, JS, and OSD to enhance integration and coordination of subordinate and supporting plans with the DOD global campaign plan.

(i) Accompanies supporting CCDRs as they brief their supporting plans through final approval, as required. To ensure coordination, all plans should be briefed at the same time.

(j) Develops assessment criteria and timelines. Collects and collates assessments, and provides feedback on plan success (e.g., accomplishment of intermediate objectives, milestones) through IPRs and the AJA process.

(k) In coordination with the JS, makes recommendations for the communication annex.

(l) The JSCP may provide additional guidance on coordinating authority based on specific planning requirements.

(4) Supporting CCDRs, Services, the National Guard Bureau, and applicable DOD agencies:

(a) Provide detailed planning support to the lead CCMD to assist in development of the DOD-wide campaign plan.

(b) Support plan conferences and planning efforts.

(c) Develop supporting plans consistent with the strategic framework, planning guidance, and process established by the lead planner.

(d) Provide subordinate or supporting plans to the lead planner prior to IPRs with enough time for the lead CCMD to review and propose modifications prior to the IPR.

g. The SecDef may also direct the CJCS to support global campaign planning. This designation will not change command relationships, but takes advantage of the CJCS's position to look across the CCMDs and provide a global perspective of opportunities and risk in developing globally integrated plans.

3. Conditions, Objectives, Effects, and Tasks Linkage

a. For CCMD campaign plans, the CCDR develops military objectives to aid in focusing the strategy and campaign plan. CCDRs' strategies establish long-range objectives to provide context for intermediate objectives. Achieving intermediate objectives sets conditions to achieve the command's objectives. The CCDR and planners update the CCMD's strategy and TCP based on changes to national objectives, achievement of TCP objectives, and changes in the OE.

b. Conditions describe the state of the OE. These are separate from the objective, as an objective may be achieved, but fail to set the desired conditions.

c. Objectives are clearly defined, measurable, and attainable. Intermediate objectives serve as waypoints against which the CCMD can measure success in attaining GEF-directed and national objectives.

d. Tasks direct friendly actions to create desired effect(s). These are the discrete activities directed in the campaign plan used to influence the OE. The execution of a task will result in an effect.

4. Resource-Informed Planning (Capability Assignment, Apportionment, Allocation)

a. GEF- and JSCP-directed campaigns, unlike contingency plans, are plans in execution. They are constrained by the readiness and availability of resources and authorities and forecast future requirements based on projected results of current on-going operations and activities.

b. CCDRs are responsible for planning, assessing, and executing their GEF- and JSCP-directed campaign plans. The CCMDs, however, receive limited budgeting and rely on the Services and the CCMD component commands to budget for and execute campaign activities. As such, the components, Joint Force Coordinator and JFPs must be involved during the planning process to identify resources and tools that are likely to be made available to ensure the campaign plan is executable. The component commands can also identify options and activities of which the CCMD might not be aware.

c. Campaign planning requires planning across four resource timeframes (see Figure III-2).

(1) Ongoing operations are executed with the current budget and assigned and currently allocated resources. As the operations progress and the CCMD conducts its assessment, the commander may be able to redesignate assigned and allocated resources, with the proper authorities, to more effective operations and activities or to address critical issues that may arise. Simultaneously, the commander uses the ongoing assessment to project a resource requirement for two years in the future (the program year). The commander uses assessment of the OE and the projection of the impact of activities in both the current and budget year (which are already locked in).

(2) The commander develops and briefs the campaign plan for the upcoming year considering the budget year forecast and force assignments and allocations, Service ceilings, and the apportionment tables. The commander updates intermediate objectives, develops new ones as appropriate, and prioritizes resources based on the ongoing assessment of current year actions. This plan is briefed through the JS to SecDef (or designated representative). The commander also identifies gaps and shortfalls in capabilities, along with associated risk, and includes them in the integrated priority list and strategic and military risk in the commander's input to the annual AJA. These reports support the command's budget and force request for the budget and apportionment in development (program year).

(3) The commander uses the current and budget year allocation, combined with the assessment, to develop a budget and resource request for the program years. Working

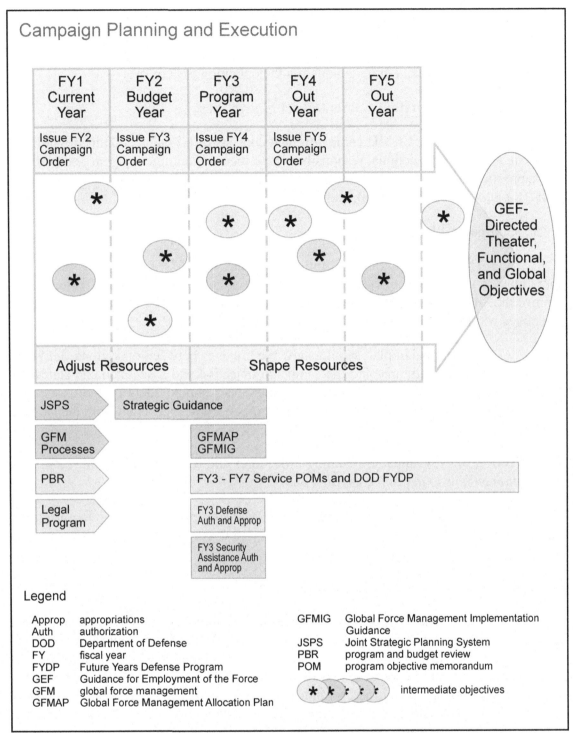

Figure III-2. Campaign Planning and Execution

with the JS, the command identifies opportunities for military engagement, exercises (joint and combined), and identifies future posture requirements that support the CCMD campaign. Posture changes, in particular, require long-lead times to implement, so the commander has to identify these in time to conduct required diplomacy and stationing requirements (such as construction) to meet any posture changes.

(4) The commander can only execute those operations and activities for which there are resources and authorities. The commander may be further restricted by the authorizations or laws that limit the use of the resources for specific programs or require specific conditions be met before conducting the operation or activity.

5. Elements of a Combatant Command Campaign Plan

a. **Overview.** The CCMD campaign plan consists of all plans contained within the established theater or functional responsibilities to include contingency plans, subordinate and supporting plans, posture plans, country-specific security cooperation sections/country plans (for geographic commands), and operations in execution.

b. **Campaign Plan**

(1) The campaign plan operationalizes the CCDR's strategy by organizing operations, activities, and investments within the assigned and allocated resources to achieve the GEF- and JSCP-directed objectives, as well as additional CCDR-determined objectives within the time frame established by the GEF or JSCP.

(2) The campaign plan should show the linkages between operations, activities, investments, and expenditures and the campaign objective and associated end states that available resources will support. The campaign plan should identify the assessment process by which the command ascertains progress toward or regression from the national security objectives.

See CJCSM 3130.01, Campaign Planning Procedures and Responsibilities, *for additional information on how to develop campaign plans.*

c. **Posture Plan.** The posture plan is the CCMD's proposal for forces, footprint, and agreements required and authorized to achieve the command's objectives and set conditions for accomplishing assigned missions.

d. **Theater Logistics and Distribution Plans**

(1) **TDP.** The TDP provides detailed theater mobility and distribution analysis to ensure sufficient capacity or planned enhanced capability throughout the theater and synchronization of distribution planning throughout the global distribution network. The TDP includes a comprehensive list of references, country data, and information requirements necessary to plan, assess, and conduct theater distribution and joint reception, staging, onward movement, and integration (JRSOI) operations. The GCCs develop their TDPs using the format in USTRANSCOM's *Campaign Plan for Global Distribution,* JSCP, and JSCP Logistics Supplement. TDPs and TPPs complement each other by posturing forces, footprints, and agreements that will interface with the GCC's theater distribution network in order to provide a continuous flow of material and equipment into the AOR. This synchronization enables a GCC's theater distribution pipeline to have sufficient capacity and capability to support development of TCPs, OPLANs, and CONPLANs.

For more information, see Appendix J, "Theater Distribution Plans."

(2) **Theater Logistics Overview (TLO).** The TLO codifies the GCC's theater logistics analysis (TLA) within the TPP. The TLO provides a narrative overview, with supporting matrices of key findings and capabilities from the TLA, which is included in the TPP as an appendix.

(3) **TLA.** The TLA provides detailed country by country analysis of key infrastructure by location or installation (main operating base [MOB]/forward operating site [FOS]/cooperative security location [CSL]), footprint projections, HN agreements, existing contracts, and task orders required to logistically support theater campaigns and their embedded contingency operations.

e. **Regional and Country-Specific Security Cooperation Sections/Country Plans**

(1) As needed or directed, CCDRs prepare country-specific security cooperation sections/country plans within their campaign plans for each country where the CCMD intends to apply significant time, money, and/or effort. CCDRs may also prepare separate regional plans. These are useful to identify and call out activities directed toward specific regional or country objectives and provide focus for the command.

(2) Regional and country-specific security cooperation sections/country plans can also serve to better harmonize activities and investments with other agencies. By isolating the desired objectives, planners can more easily identify supporting efforts and specific assessment measures toward achieving US objectives.

(3) Where the US has identified specific objectives with a country or region (through strategic guidance or policy), separate regional or country-specific security cooperation sections/country plans help to identify resource requirements and risk associated with resource limitations that may be imposed.

f. **Subordinate, Supporting, and Campaign Support Plans**

(1) **Subordinate Campaign Plan.** JFCs subordinate to a CCDR or other JFC may develop subordinate campaign plans in support of the higher plan to better synchronize operations in time and space. It may, depending upon the circumstances, transition to a supported or supporting plan in execution.

(2) Supporting plans are prepared by a supporting commander, a subordinate commander, or the head of a department or agency to satisfy the requests or requirements of the supported commander's plan.

(3) Campaign support plans are developed by the Services, National Guard Bureau, and DOD agencies that integrate the appropriate USG activities and programs, describe how they will support the CCMD campaigns, and articulate institutional or component-specific guidance.

g. **Contingency Plans.** Contingency plans are branch plans to the campaign plan that are based upon hypothetical situations for designated threats, catastrophic events, and contingent missions outside of crisis conditions. The campaign plan should address those known issues in the contingencies that can be addressed prior to execution to establish conditions, conduct deterrence, or address assumptions. As planners develop contingency plans, issues and concerns in the contingency should be included as an element of the campaign. See Chapter II, "Strategic Guidance and Coordination," subparagraph 15a(2)(e) for discussion of the levels of contingency plans.

6. Assessing Theater and Functional Campaign Plans

a. Campaign plan assessments determine the progress toward accomplishing a task, creating a condition, or achieving an objective. Campaign assessments enable the CCDR and supporting organizations to refine or adapt the campaign plan and supporting plans to achieve the campaign objectives or, with SecDef approval, to adapt the GEF-directed objectives to changes in the strategic and OEs.

b. The campaign assessment is also DOD's bridging mechanism from the CCDR's strategy to the strategic, resource, and authorities planning processes, informing DOD's strategic direction; assignment of roles and missions; and force employment, force posture, force management, and force development decision making. Through the AJA, the campaign assessment also informs the CJCS's risk assessment and the SecDef's risk mitigation plan.

c. The campaign assessment provides the CCDR's input to DOD on the capabilities needed to accomplish the missions in the contingency plans of their commands over the planning horizon of the CCDR's strategy, taking into account expected changes in threats and the strategic and OEs.

d. Assessments enable the CCDR to make the case for additional resources or to recommend re-allocating available resources to the highest priorities. The assessment allows SecDef and senior leaders to do the same across all CCMDs and to make the case to Congress to add or re-allocate resources through the Future Years Defense Program (FYDP).

7. Risk

a. GCCs assess how strongly US interests are held within their respective areas, how those interests can be threatened, and their ability to execute assigned missions to protect them and achieve US national objectives. This is documented in the CCDR's strategic estimate and in the annual submission to the AJA.

b. CCDRs and DOD's senior leaders work together to reach a common understanding of integrated risk (the strategic risk assessed at the CCMD level combined with the military risk), decide what risk is acceptable, and minimize the effects of accepted risk by establishing appropriate risk controls.

c. For strategic risk, CCDRs identify the probability and consequence of near (0-2 years) and mid-term (3-7 years) strategic events or crises that could harm US national interests, and they identify the impacts of long-term (8-20 years) trends and future adversary capabilities.

d. For military risk, CCMDs evaluate the impact of the difference between required and available capability, capacity, readiness, plans, and authorities on their ability to execute assigned missions. Military risk is composed of the risk to mission assessed by the CCMD, risk to the force assessed by the Services, and risk to potential future operations. Assessments include, but are not limited to:

(1) FYDP budgetary priorities, tradeoffs, or fiscal constraints.

(2) Deficiencies and strengths in force capabilities identified during preparation and review of campaign and contingency plans.

(3) Projected readiness of forces required to execute the campaign in future years.

(4) Assumptions or plans about contributions or support of:

(a) Other USG departments and agencies.

(b) Alliances, coalitions, and other friendly nations.

(c) Operational contract support (OCS).

(d) Changes in adversary capabilities identified during the preparation of the strategic estimate and other intelligence products.

e. Commanders must be willing to stop unproductive and minimally productive activities. Although there is currently no proven cost-benefit analysis for strategic assessment, the commanders should be willing to try new activities to see if there are better or less risky methods to achieve theater and national objectives.

For additional information on risk, see CJCSM 3105.01, Joint Risk Analysis.

8. Opportunity

a. CCDRs need to identify opportunities they can exploit to influence the situation in a positive direction. Limited windows of opportunity may open and the CCDR must be ready to exploit these to set the conditions that will lead to successful transformation of the conflict and thus to transition. This should be done in collaboration with interagency partners, international partners, and partner nations who may have assessment tools that look for opportunities to enhance resilience and mitigate conflict.

b. It is important to comprehend dynamics such as evolving strategic guidance and mandates, the type of conflict, the strategic logic of perpetrators, the impact of operations, and changing vulnerabilities and threats that relate to protection of civilians, resiliencies, and emerging opportunities, to enhance positive changes in the OE or among the actors.

c. Assessing the OE from the perspective of the root causes and immediate drivers of instability is essential to identify and create opportunities for longer-term processes to deal with the root causes.

d. Successful conflict transformation relies on the ability of the joint force along with the other intervening actors and local stakeholders to identify and resolve the primary sources of instability by focusing on the underlying sources of that instability, while also managing its visible symptoms. In countries seeking to transition from war to peace, a limited window of opportunity exists to mitigate sources of instability. This may include deterring adversaries and mitigating their effects on local populaces and institutions, as well as developing approaches that include marginalized groups, consensus-building mechanisms, checks and balances on power, and transparency measures.

For more information on root causes and drivers of conflict, see JP 3-07, Stability.

CHAPTER IV
OPERATIONAL ART AND OPERATIONAL DESIGN

> *"War is an art and as such is not susceptible of explanation by fixed formula."*
>
> **General George S. Patton, Jr., Success in War, *The Infantry Journal Reader*,**
> **1931**

1. Overview

a. The JFC and staff develop plans and orders through the application of operational art and operational design in conjunction with JPP. They combine art and science to develop products that describe how (ways) the joint force will employ its capabilities (means) to achieve military objectives (ends), given an understanding of unacceptable consequences of employing capabilities as intended (risk).

(1) Operational art is the cognitive approach by commanders and staffs—supported by their skill, knowledge, experience, creativity, and judgment—to develop strategies, campaigns, and operations to organize and employ military forces by integrating ends, ways, means, and risks. Operational art is inherent in all aspects of operational design.

(2) Operational design is the conception and construction of the framework that underpins a campaign or operation and its subsequent execution. The framework is built upon an iterative process that creates a shared understanding of the OE; identifies and frames problems within that OE; and develops approaches, through the application of operational art, to resolving those problems, consistent with strategic guidance and/or policy. The **operational approach,** a primary product of operational design, allows the commander to continue JPP, translating broad strategic and operational concepts into specific missions and tasks (see Figure IV-1) to produce an executable plan.

b. **The purpose of operational design and operational art is to produce an operational approach,** allowing the commander to continue JPP, translating broad strategic and operational concepts into specific missions and tasks and produce an executable plan.

c. Operational design is one of several tools available to help the JFC and staff understand the OE and develop broad solutions for mission accomplishment and understand the uncertainty in a complex OE. Additionally, it supports a recursive and ongoing dialogue concerning the nature of the problem and an operational approach to achieve the desired objectives.

d. Operational design and operational art enable **understanding.** Understanding is more than just knowledge of the capabilities and capacities of the relevant actors (individuals and organizations) or the nature of the OE, it provides context for decision making and how the many facets of the problem are likely to interact, allowing commanders and planners to identify consequences, opportunities, and recognize risk. The

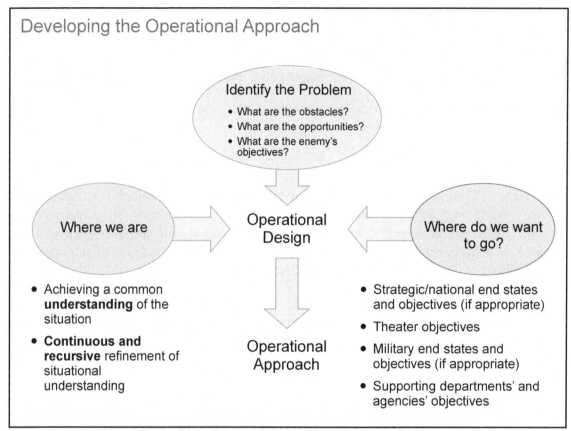

Figure IV-1. Developing the Operational Approach

tools described in this chapter are meant to aid commanders in conducting robust analysis, particularly in handling unexpected events or those events outside of their previous experience or understanding. Robust analysis will aid in better understanding and ultimately better decision making.

e. Implementation is based on the commander's and planners' experience and time available. Different commanders and planners will need different tools to help them as each person has inherent strengths, weaknesses, and prejudices. Similarly, every problem is different and may require different tools to analyze and address it. The tools chosen by the planner should be appropriate for the problem and should complement the planners' strengths and weaknesses.

(1) The amount of data readily available today can quickly overwhelm the planning process. Planners and commanders need to understand that a good timely decision with incomplete information may present a better solution than waiting until all information is available.

(2) In the complex social systems that are an integral part of military operations, additional data can greatly increase the complexity of the problem without aiding understanding. Operational art aids the commander in identifying the point of diminishing returns in collection and analysis.

2. The Commander's Role

a. **The commander is the central figure in operational design** due to knowledge and experience, and because the commander's judgment and decisions are required to guide the staff through the process. Generally, the more complex a situation, the more critical the role of the commander early in planning. Commanders draw on operational design to mitigate the challenges of complexity and uncertainty, as well as leveraging their knowledge, experience, judgment, intuition, responsibility, and authority to generate a clearer understanding of the conditions needed to focus effort and achieve success.

b. Commanders distinguish the unique features of their current situations to enable development of innovative or adaptive solutions. They understand that each situation requires a solution tailored to the context of the problem. Through the use of operational design and the application of operational art, commanders develop innovative, adaptive alternatives to solve complex challenges. These broad alternatives are the operational approach (Figure IV-1).

c. Commanders use the knowledge and understanding gained from operational design, along with any additional guidance from higher headquarters, to provide commander's guidance that directs and guides the staff through JPP in preparing detailed plans and orders. Developing meaningful touch-points throughout the planning process with the supported and supporting commanders and other stakeholders enables a shared understanding of the OE.

d. Operational design requires the commander to encourage discourse and leverage dialogue and collaboration to identify complex, ill-defined problems. To that end, the commander must empower organizational learning and develop methods to determine whether modifying the operational approach is necessary during the course of an operation or campaign. This requires assessment and reflection that challenge understanding of the existing problem and the relevance of actions addressing that problem. Due to complexity and constant change, commanders should be comfortable in the recognition that they will never know everything about the given OE and will never be able to fully define its problems. As such, many of the problems in the OE may not have solutions.

e. **Red Teaming**

(1) Gathering and analyzing information—along with discerning the perceptions of adversaries, enemies, partners, and other relevant actors—is necessary to correctly frame the problem, which enables planning. A red team, an independent group that challenges an organization to improve its effectiveness, can aid a commander and the staff to think critically and creatively; see things from varying perspectives; challenge their thinking; avoid false mind-sets, biases, or group thinking; and avoid the use of inaccurate analogies to frame the problem.

(2) Red teaming provides an independent capability to fully explore alternatives in plans and operations in the context of the OE and from the perspective of adversaries and other relevant actors.

(3) Commanders use red teams to aid them and their staffs to provide insights and alternatives during planning, execution, and assessment to:

(a) Broaden the understanding of the OE.

(b) Assist the commander and staff in framing problems and defining end state conditions.

(c) Challenge assumptions.

(d) Consider the perspectives of the adversary and other relevant actors as appropriate.

(e) Aid in identifying friendly and enemy vulnerabilities and opportunities.

(f) Assist in identifying areas for assessment as well as the assessment metrics.

(g) Anticipate the cultural perceptions of partners, adversaries, and other relevant actors.

(h) Conduct independent critical reviews and analyses of plans to identify potential weaknesses and vulnerabilities.

(4) Red teams provide the commander and staff with an independent capability to challenge the organization's thinking.

(5) The red team crosses staff functions and time horizons in JPP, which is different than a red cell, which is composed of members of the intelligence directorate of a joint staff (J-2) and performs threat emulation, or a joint intelligence operations center (JIOC) red team as an additive element on the J-2 staff to improve the intelligence analysis, products, and processes.

For more discussion on red teams, see Appendix K, "Red Teams."

SECTION A. OPERATIONAL ART

3. Overview

a. Commanders, skilled in the use of operational art, provide the vision that links strategic objectives to tactical tasks through their understanding of the strategic and OEs during both the planning and execution phases of an operation or campaign. More specifically, the interaction of operational art and operational design provides a bridge between strategy and tactics, linking national strategic aims to operations that must be executed to accomplish these aims and identifying how to assess the impact of the operations in achieving the strategic objectives. Likewise, operational art promotes unified action by helping JFCs and staffs understand how to facilitate the integration of other agencies and multinational partners toward achieving strategic and operational objectives.

b. Through operational art, commanders link ends, ways, and means to attain the desired end state (see Figure IV-2). This requires commanders to answer the following questions:

(1) What is the current state of the OE?

(2) What are the military objectives that must be achieved, how are they related to the strategic objectives, and what objectives must be achieved to enable that strategic/national objective? How do these differ from the current conditions (state of the OE)? **(Ends)**

(3) What sequence of military actions, in conjunction with possible civilian actions, is most likely to achieve those objectives and attain the end state? How will I measure achievement of those objectives? **(Ways)**

(4) What military resources are required in concert with possible civilian resources to accomplish that sequence of actions within given or requested resources? **(Means)**

(5) What is the chance of failure or unacceptable consequences in performing that sequence of military actions? How will I identify if one or more of them occur? What is an acceptable level of "failure"? **(Risk)**

4. Role of Operational Art

a. Operational art enables commanders and staffs to take large amounts of data generated in the planning and analysis processes and distill it into useable information. During the plan development phase, detailed analysis may be required to determine feasible approaches and identify risk. Often during the decision-making process (and in IPRs), there is insufficient time to delve into the detail used to arrive at the proposed recommendation.

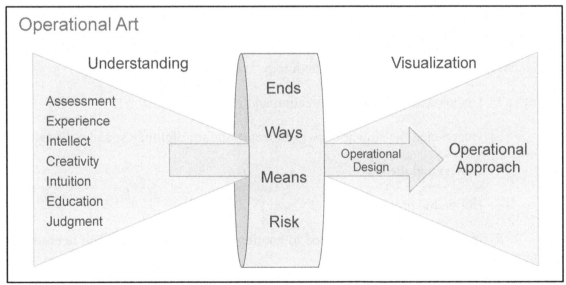

Figure IV-2. Operational Art

(1) Operational art provides the ability to better understand the OE, understand the decision-making process, and provide a concise and sufficiently detailed explanation without getting lost in the minutiae.

(2) It also provides the commander the ability to make judgments and decisions with incomplete information. This is critical in crisis planning, time-constrained planning, and during execution, when there may not be the amount of time or analytic capability desired to conduct a full analysis of the OE.

b. Operational art also provides awareness of personal and organizational biases that could affect the analysis and decision processes. Although it is often difficult to completely ignore the biases, it enables an understanding of how they affect the decision process and risk associated with those decisions.

SECTION B. OPERATIONAL DESIGN

5. Overview

a. **Operational design** is a methodology to aid commanders and planners in organizing and understanding the OE.

b. There are four major components to operational design (see Figure IV-3). The components have characteristics that exist outside of each other and are not necessarily sequential. However, an understanding of the OE and problem must be established prior to developing operational approaches.

c. Operational design is one of several tools available to help the JFC and staff understand the broad solutions for mission accomplishment and to understand the uncertainty in a complex OE. Additionally, it supports a recursive and ongoing dialogue concerning the nature of the problem and an operational approach to achieve the desired objectives.

d. The process is continuous and cyclical in that it is conducted prior to, during, and for follow-on joint operations.

e. **Methodology.** The general methodology in operational design is:

(1) Understand the strategic direction and guidance.

(2) Understand the strategic environment (policies, diplomacy, and politics).

(3) Understand the OE.

(4) Define the problem.

(5) Identify assumptions needed to continue planning (strategic and operational assumptions).

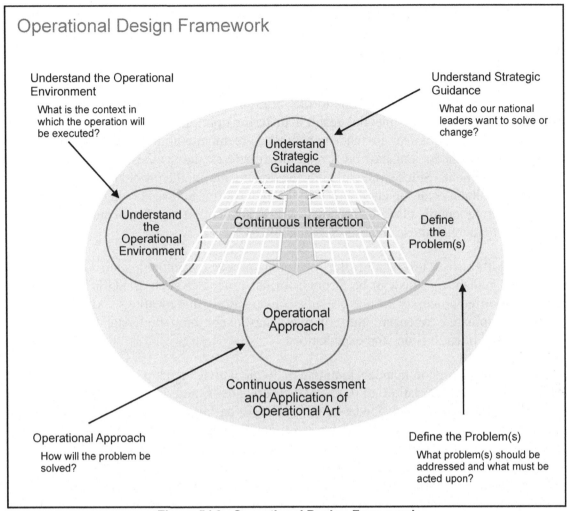

Figure IV-3. Operational Design Framework

    (6) Develop options (the operational approach).

    (7) Identify decisions and decision points (external to the organization).

    (8) Refine the operational approach(es).

    (9) Develop planning guidance.

    f. These steps are not necessarily sequential. Iteration and reexamination of earlier work is essential to identify how later decisions affect earlier assumptions and to fill in gaps identified during the process.

6. Understand the Strategic Direction and Guidance

    a. Planning usually starts with the assignment of a planning task through a directive, order, or cyclical strategic guidance depending on how a situation develops. The commander and staff must analyze all available sources of guidance. These sources include written documents, such as the GEF and JSCP, written directives, oral instructions from higher headquarters, domestic and international laws, policies of other organizations that are interested in the situation, communication synchronization guidance, and higher headquarters' orders or estimates.

b. Strategic direction from strategic guidance documents can be vague, incomplete, outdated, or conflicting. This is due to the different times at which they may have been produced, changes in personnel that result in differing opinions or policies, and the staffing process where compromises are made to achieve agreement within the documents. During planning, commanders and staff must read the directives and synthesize the contents into a concise statement. Since strategic guidance documents can be problematic, the JFC and staff should obtain clear, updated direction through routine and sustained civilian-military dialogue throughout the planning process. When clarification does not occur, planners and commanders identify those areas as elements of risk.

c. Additionally, throughout the planning process, senior leaders will provide additional guidance. This can be through formal processes such as SGSs and IPRs, or through informal processes such as e-mails, conversations, and meetings. All of this needs to be disseminated to ensure the command has a common understanding of higher commander's intent, vision, and expectations.

d. In particular, commanders maintain dialogue with leadership at all levels to resolve differences of interpretation of higher-level objectives and the ways and means to accomplish these objectives. Understanding the OE, defining the problem, and devising a sound approach, are rarely achieved the first time. Strategic guidance addressing complex problems can initially be vague, requiring the commander to interpret and filter it for the staff. While CCDRs and national leaders may have a clear strategic perspective of the problem from their vantage point, operational-level commanders and subordinate leaders often have a better understanding of specific circumstances that comprise the operational situation and may have a completely different perspective on the causes and solutions. Both perspectives are essential to a sound plan. Subordinate commanders should be aggressive in sharing their perspective with their higher headquarters, and both should resolve differences at the earliest opportunity. While policy and strategic guidance clarify planning, it is equally true that planning informs policy formulation. A strategy or plan that cannot be realistically executed at the tactical level can be as detrimental to long-range US interests as tactical actions that accomplish a task but undermine the strategic or operational objectives.

e. Strategic guidance is essential to operational art and operational design. As discussed in Chapter I, "Joint Planning," the President, SecDef, and CJCS all promulgate strategic guidance. In general, this guidance provides long-term as well as intermediate or ancillary objectives. It should define what constitutes victory or success **(ends)** and identify available forces, resources, and authorities **(means)** to achieve strategic objectives. The operational approach **(ways)** of employing military capabilities to achieve the ends is for the supported JFC to develop and propose, although policy or national positions may limit options available to the commander. Connecting resources and tactical actions to strategic ends is the responsibility of the operational commander—the commander must be able to explain how proposed actions will result in desired effects, as well as the potential risks of such actions.

f. For situations that require the employment of military capabilities (particularly for anticipated large-scale combat), the President and SecDef may establish a set of **operational objectives.** However, in the absence of coherent guidance or direction, the CCDR/JFC may need to collaborate with policymakers in the development of these objectives. Achievement of these objectives should result in contributing to the **strategic objective—the broadly expressed conditions that should exist after the conclusion of a campaign or operation.** Based on the ongoing civilian-military dialogue, the CCDR will determine the **military end state** and military objectives, which define the role of military forces. These objectives are the basis for operational design.

7. Understand the Strategic Environment

a. After analyzing the strategic guidance, commanders and planners build an understanding of the strategic environment. This forms boundaries within which the operational approach must fit. Some considerations are:

(1) What actions or planning assumptions will be acceptable given the current US policies and the diplomatic and political environment?

(2) What impact will US activities have on third parties (focus on military impacts but identify possible political fallout)?

(3) What are the current national strategic objectives of the USG? Are the objectives expected to be long lasting or short-term only? Could they result in unintended consequences (e.g., if you provide weapons to a nation, is there sufficient time to develop strong controls so the weapons will not be used for unintended purposes)?

b. **Strategic-Level Considerations.** Strategic-level military activities affect national and multinational military objectives, develop CCMD campaign plans to achieve these objectives, sequence military operations, define limits and assess risks for use of the military instrument of national power, and provide military forces and capabilities in accordance with authorizing directives. Within the OE, there are strategic-level considerations that may include global aspects due to global factors such as international law, the capability of adversary/enemy information activities to influence world opinion, adversary and friendly organizations and institutions, and the capability and availability of national and commercial space-based systems and information technology. Strategic-level considerations of the OE are analyzed in terms of geopolitical regions, nations, and climate rather than local geography and weather. Nonmilitary aspects of the OE assume increased importance at the strategic level. For example, the industrial and technological capabilities of a nation or region will influence the type of military force it fields, and factors may influence the ability of a nation or region to endure a protracted conflict without outside assistance. In many situations, nonmilitary considerations may play a greater role than military factors in influencing adversary and relevant actor COAs. The JIPOE process analyzes all relevant aspects of the OE, including the adversary and other actors, and

PMESII systems and subsystems. This analysis should also consider possible intervention by third parties. The main JIPOE focus is to provide predictive intelligence that helps the JFC discern the adversary's probable intent and most likely future COA. During COA development, analysis, comparison, and approval during JPP, JIPOE-based COA models consider the entire range of resources available to the adversary, to include the mindset of key personalities and populations, and the financial flows and convergence of threat and illicit networks to fund adversary operations. JIPOE-based COA models identify both military and nonmilitary methods of power projection and influence, specify the theaters of main effort and the forces committed to each, and depict national as well as strategic and theater-level objectives of the relevant actors.

8. Understand the Operational Environment

a. The OE is the composite of the conditions, circumstances, and influences that affect the employment of capabilities and bear on the decisions of the commander. It encompasses physical areas and factors of the air, land, maritime, and space domains; the electromagnetic spectrum; and the information environment (which includes cyberspace). Included within these areas are the adversary, friendly, and neutral actors that are relevant to a specific joint operation. Understanding the OE helps the JFC to better identify the problem; anticipate potential outcomes; and understand the results of various friendly, adversary, and neutral actions and how these actions affect attaining the military end state (see Figure IV-4).

b. The commander must be able to describe both the current state of the OE and the desired state of the OE when operations conclude (desired military end state) to visualize an approach to solving the problem. Planners can compare the current conditions of the OE with the desired conditions. Identifying necessary objective conditions and termination criteria early in planning will help the commander and staff devise an operational approach with LOEs/LOOs that link each current condition to a desired end state condition.

c. **Describe the Current OE.** The JIPOE process is a comprehensive analytic tool to describe all aspects of the OE relevant to the operation or campaign.

d. **Operational-Level Considerations**

(1) In analyzing the current and future OE, the staff can use a PMESII analytical framework to determine relationships and interdependencies relevant to the specific operation or campaign (see Figure IV-5).

(2) The size and scope of the analysis may also vary depending on particular aspects of the OE. For example, if a landlocked adversary has the capability to conduct space-based intelligence collection or cyberspace operations, then the relevant portions of space and the information environment would extend worldwide, while maritime considerations might be minimal. While most joint operations at the operational level may encompass many or all PMESII considerations and characteristics, the staff's balanced JIPOE efforts should vary according to the relevant OE aspects of the operation or campaign.

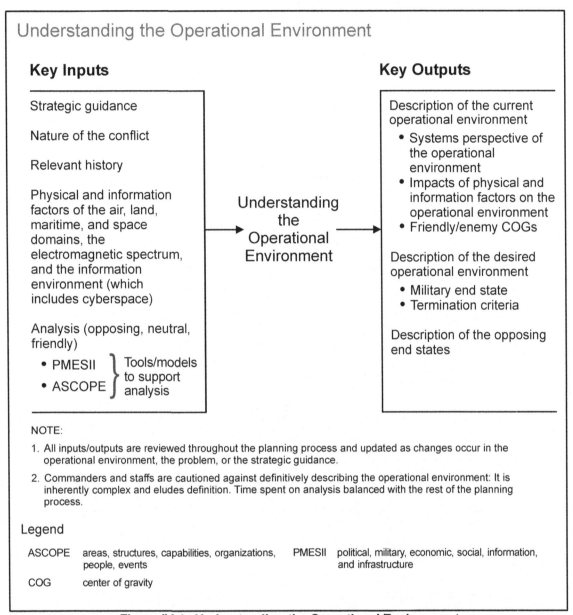

Figure IV-4. Understanding the Operational Environment

See JP 2-01.3, Joint Intelligence Preparation of the Operational Environment, *for additional information on analyzing and understanding the OE.*

    (3) Additional factors that should be considered, include:

        (a) Geographical features and meteorological and oceanographic characteristics.

        (b) Population demographics (ethnic groups, tribes, ideological factions, religious groups and sects, language dialects, age distribution, income groups, public health issues).

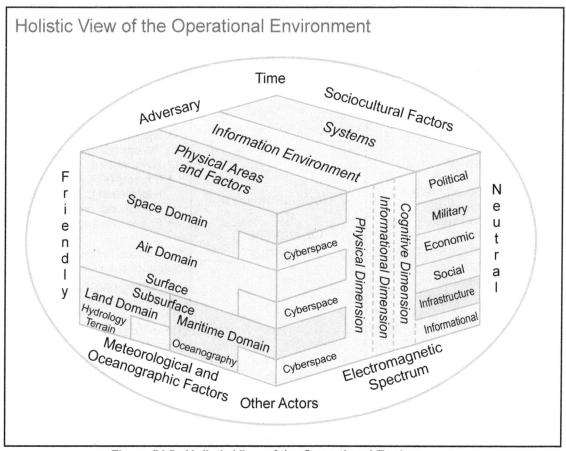

Figure IV-5. Holistic View of the Operational Environment

(c) Social and cultural factors of adversaries, neutrals, and allies in the OE (beliefs, how and where they get their information, types and locations of media outlets).

(d) Political and socioeconomic factors (economic system, political factions, tribal factions).

(e) Infrastructure, such as transportation, energy, and information systems.

(f) Operational limitations such as rules of engagement (ROE), rules for the use of force (RUF), or legal restrictions on military operations as specified in US law, international law, or HN agreements.

(g) All friendly, adversary, and enemy conventional, irregular, and paramilitary forces and their general capabilities and strategic objectives (including all known and/or suspected chemical, biological, radiological, and nuclear threats and hazards).

(h) Environmental conditions (earthquakes, volcanic activity, pollution, naturally occurring diseases).

(i) Location of toxic industrial materials in the area of interest that may produce chemical, biological, radiological, or nuclear hazards.

(j) Psychological characteristics of adversary decision making.

(k) All locations of foreign embassies, international organizations, and NGOs.

(l) Friendly and adversary military and commercial capabilities provided by assets in space, their current or potential use, and critical vulnerabilities.

(m) Knowledge of the capabilities and intent, COGs, and critical vulnerabilities of forces, individuals, or organizations conducting cyberspace operations.

(n) Financial networks that could impact the adversary's ability to sustain operations.

(4) To produce a holistic view of the relevant adversary, neutral, and friendly systems within a larger system that includes many external influences, analysis should define how these systems interrelate. Most important to this analysis is describing the **relevant relationships** within and between the various systems that directly or indirectly affect the problem at hand. Although the J-2 manages the JIPOE process, other directorates and agencies can contribute valuable expertise to develop and assess the complexities of the OE.

For more information on JIPOE, see JP 2-01.3, Joint Intelligence Preparation of the Operational Environment.

(5) **Tendencies and Potentials.** In developing an understanding of the interactions and relationships of relevant actors in the OE, commanders and staffs consider observed tendencies and potentials in their analyses. Tendencies reflect the inclination to think or behave in a certain manner. Tendencies are not considered deterministic but rather model the thoughts or behaviors of relevant actors. Tendencies help identify the range of possibilities that relevant actors may develop with or without external influence. Once identified, commanders and staffs evaluate the potential of these tendencies to manifest within the OE. Potential is the inherent ability or capacity for the growth or development of a specific interaction or relationship. However, not all interactions and relationships support attaining the desired end state. The desired end state accounts for tendencies and potentials that exist among the relevant actors or other aspects of the OE. Early in JPP, pertinent lessons learned should be collected and reviewed as part of the analysis to allow previously learned lessons to make their way into the plan. The Joint Lessons Learned Information System provides a database of past lessons learned. However, people experienced in the mission, OE, and lessons learned functions should be sought for their knowledge and experience.

(6) Describe the key conditions that must exist in the future OE to achieve the objectives. Planners should put a temporal aspect to this set of conditions in order to be able to conduct feasibility and acceptability analyses.

(7) Determine the objectives of relevant actors affecting the OE. These actors will have different sets of conditions for achieving their respective objectives. Such

opposition can be expected to take actions to thwart US and partner nations' objectives. Other actors, neutral or friendly, may not have an opposing mindset, but may have desired conditions (or unintended consequences of their actions) that oppose our desired end state conditions. The analysis of the OE should identify where the contradictions between the competing sides, allies, partners, and neutrals lie and recognize the conflicts of interests. In the course of developing the plan, planners should ask themselves if the COA being considered addresses these conflicts.

9. Define the Problem

a. Defining the problem is essential to addressing the problem. It involves understanding and isolating the root causes of the issue at hand—defining the essence of a complex, ill-defined problem. Defining the problem begins with a review of the tendencies and potentials of the relevant actors and identifying the relationships and interactions among their respective desired conditions and objectives. The problem statement articulates how the operational variables can be expected to resist or facilitate transformation and how inertia in the OE can be leveraged to ensure the desired conditions are achieved.

(1) The problem statement identifies the areas for action that will transform existing conditions toward the desired end state. Defining the problem extends beyond analyzing interactions and relationships in the OE (see Figure IV-6). It identifies areas of tension and competition—as well as opportunities and challenges—that commanders must address to transform current conditions to attain the desired end state. Tension is the resistance or friction among and between actors. The commander and staff identify the tension by analyzing the context of the relevant actors' tendencies and potentials within the complex systems within the OE.

(2) Critical to defining the problem is determining what needs to be acted on to reconcile the differences between existing and desired conditions. Some of the conditions are critical to success, some are not. Some may be achieved as a secondary or tertiary result of another condition. In identifying the problem, the planning team identifies the tensions between the desired conditions and identifies the areas of tension that merit further consideration as areas of possible intervention.

(3) The JFC and staff must identify and articulate:

(a) Tensions between current conditions and desired conditions at the end state.

(b) Elements within the OE which must change or remain the same to attain desired end states.

(c) Opportunities and threats that either can be exploited or will impede the JFC from attaining the desired end state.

(d) Operational limitations.

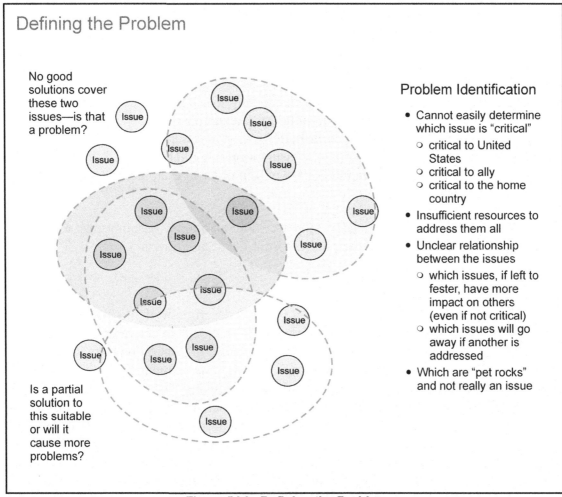

Figure IV-6. Defining the Problem

b. A concise problem statement is used to clearly define the problem or problem set to solve. It considers how tension and competition affect the OE by identifying how to transform the current conditions to the desired end state—before adversaries begin to transform current conditions to their desired end state. The statement broadly describes the requirements for transformation, anticipating changes in the OE while identifying critical transitions.

10. Identify Assumptions

a. Where there is insufficient information or guidance, the commander and staff identify assumptions to assist in framing solutions. At this stage, assumptions address strategic and operational gaps that enable the commander to develop the operational approach.

(1) Assumptions should be kept to the minimum required as each assumption adds to the probability of error in the plan and requires specific CCIRs to continuously check its validity.

(2) Assumptions address key and critical decisions required by senior leaders to enable the continuation of planning.

b. Commanders and staff should review strategic guidance and direction to see if any assumptions are imposed on the planning process. They should also regularly discuss planning assumptions with OSD and DOD leadership to see if there are changes in policy or guidance that affect the planning assumptions (examples could be basing or access permissions, allied or multinational contributions, alert and warning decision timelines, or anticipated threat actions and reactions). Assumptions should be phrased in terms of will or will not (rather than using "should" or "may") in order to establish specific conditions that enable planning to continue.

c. During JPP, the commander may develop additional assumptions to support detailed COA development (see Chapter V, "Joint Planning Process").

11. Developing Operational Approaches

a. The operational approach is a commander's description of the broad actions the force can take to achieve an objective in support of the national objective or attain a military end state. It is the commander's visualization of how the operation should transform current conditions into the desired conditions—the way the commander envisions the OE at the conclusion of operations to support national objectives. The operational approach is based largely on an understanding of the OE and the problem facing the JFC. A discussion of operational approaches within and between options forms the basis of the IPRs between the CCDR and SecDef and staff (to ensure consistency with US policy and national objectives). Once SecDef approves the approach, it provides the basis for beginning, continuing, or completing detailed planning. The JFC and staff should continually review, update, and modify the approach as policy, the OE, end states, or the problem change. This requires frequent and continuing dialogue at all levels of command.

b. Commanders and their staffs can use operational design when planning any joint campaign or operation. Notwithstanding a commander's judgment, education, and experience, the OE often presents situations so complex that understanding them—let alone attempting to change them—exceeds individual capacity. Nor does such complexity lend itself to coherent planning. Bringing adequate order to complex problems to facilitate further detailed planning requires an iterative dialogue between commander, the planning staff, and policy staff. Rarely will members of either staff recognize an implicit operational approach during their initial analysis and synthesis of the OE. Successful development of the approach requires continuous analysis, learning, dialogue, and collaboration between commander and staff, as well as other subject matter experts. The challenge is even greater when the joint operation involves other agencies, the private sector, and multinational partners (which is typically the case), whose unique considerations can complicate the problem.

c. It is essential that commanders, through a dialogue with their staffs, planning teams, initiative groups, and any other relevant sources of information, first gain an understanding of the OE, to include the US policy perspective, and define the problem facing the joint force prior to conducting detailed planning. The problem as presented in guidance documents rarely includes all available guidance information and may identify the symptoms rather than the actual problem. From this understanding of the OE and definition of the problem, commanders develop their broad operational approach for transforming current conditions into desired conditions. The operational approach will underpin the operation and the detailed planning that follows. As detailed planning occurs, the JFC and staff continue discourse and refine their operational approach.

For additional information on assessment, see Appendix D, "Operation Assessment Plan (Examples)."

12. Identify Decisions and Decision Points

a. During planning, commanders inform leadership of the decisions that will need to be made, when they will have to be made, and the uncertainty and risk accompanying decisions and delay. This provides leaders, both military and civilian, a template and warning for the decisions in advance and provides them the opportunity to look across interagency partners and with allies to look for alternatives and opportunities short of escalation. The decision matrix also identifies the expected indicators needed in support of the intelligence collection plan.

b. Commanders are responsible to ensure senior leaders understand the risk and time lines associated with the decision points and the possible effects of delayed decisions.

13. Refine the Operational Approach

a. Throughout the planning processes, commanders and their staffs conduct formal and informal discussions at all levels of the chain of command. These discussions help refine assumptions, limitations, and decision points that could affect the operational approach and ensure the plan remains feasible, acceptable, and adequate.

b. The commander adjusts the operational approach based on feedback from the formal and informal discussions at all levels of command and other information.

14. Prepare Planning Guidance

a. **Developing Commander's Planning Guidance.** The commander provides a summary of the OE and the problem, along with a visualization of the operational approach, to the staff and to other partners through commander's planning guidance. As time permits, the commander may have been able to apply operational design to think through the campaign or operation before the staff begins JPP. In this case, the commander provides initial planning guidance to help focus the staff in mission analysis. Commanders should continue the analysis to further understand and visualize the OE as the staff conducts mission analysis. Upon completing analysis of the OE, the commander will issue planning guidance, as appropriate, to help focus the staff efforts. At a minimum, the commander issues planning guidance, either initial or refined, at the conclusion of mission analysis, and provides refined planning guidance as understanding of the OE, the problem, and visualization of the operational approach matures. It is critical for the commander to provide updated guidance as the campaign or operation develops in order to adapt the operational approach to a changing OE or changed problem.

b. The format for the commander's planning guidance varies based on the personality of the commander and the level of command, but should adequately describe the logic to the commander's understanding of the OE, the methodology for reaching the understanding of the problem, and a coherent description of the operational approach. It may include the following elements:

(1) **Describe the OE.** Some combination of graphics showing key relationships and tensions and a narrative describing the OE will help convey the commander's understanding to the staff and other partners.

(2) **Define the problem to be solved.** A narrative problem statement that includes a timeframe to solve the problem will best convey the commander's understanding of the problem.

(3) **Describe the operational approach.** A combination of a narrative describing objectives, decisive points, and potential LOEs and LOOs, with a summary of limitations (constraints and restraints) and risk (what can be accepted and what cannot be accepted) will help describe the operational approach.

(4) **Provide the commander's initial intent.** The commander should also include the initial intent in planning guidance. The commander's initial intent describes the purpose of the operations, desired strategic end state, military end state, and operational risks associated with the campaign or operation. It also includes where the commander will and will not accept risk during the operation. It organizes (prioritizes) desired conditions and the combinations of potential actions in time, space, and purpose. The JFC should envision and articulate how military power and joint operations, integrated with other applicable instruments of national power, will achieve strategic success, and how the command intends to measure the progress and success of its military actions and activities. It should help staff and subordinate commanders understand the intent for unified action using interorganizational coordination among all partners and participants. Through commander's intent, the commander identifies the major unifying efforts during the campaign or operation, the points and events where operations must succeed to control or establish conditions in the OE, and where other instruments of national power will play a central role. The intent must allow for decentralized execution. It provides focus to the staff and helps subordinate and supporting commanders take actions to achieve the military objectives or attain the end state without further orders, even when operations do not unfold or result as planned. While there is no specified joint format for the commander's intent, a generally accepted construct includes the purpose, end state, and risk.

(a) **Purpose.** Purpose delineates reason for the military action with respect to the mission of the next higher echelon. The purpose explains why the military action is being conducted. The purpose helps the force pursue the mission without further orders, even when actions do not unfold as planned. Thus, if an unanticipated situation arises, participating commanders understand the purpose of the forthcoming action well enough to act decisively and within the bounds of the higher commander's intent.

(b) **End State.** An end state is the set of required conditions that defines achievement of the commander's objectives. This describes what the commander desires in military end state conditions that define mission success by friendly forces. It also describes the strategic objectives and higher command's military end state and describes how reaching the JFC's military end state supports higher headquarters' end state (or national objectives).

(c) **Risk.** Defines aspects of the campaign or operation in which the commander will accept risk in lower or partial achievement or temporary conditions. It also describes areas in which it is not acceptable to accept such lower or intermediate conditions.

(d) The intent may also include operational objectives, method, and effects guidance.

(e) The commander may provide additional planning guidance such as information management, resources, or specific effects that must be created or avoided.

SECTION C. ELEMENTS OF OPERATIONAL DESIGN

15. Elements of Operational Design

The elements of operational design (Figure IV-7) can be used for all military planning. However, not all of the elements of operational design may be required for all plans.

a. **Termination**

(1) Termination criteria are the specified standards approved by the President and/or SecDef that must be met before military operations can be concluded. Termination criteria are a key element in establishing a military end state. Termination criteria describe the conditions that must exist in the OE at the cessation of military operations. The conditions must be achievable and measurable so the commander can clearly identify the achievement of the military end state. Effective planning cannot occur without a clear understanding of the military end state and the conditions that must exist to end military

Elements of Operational Design

- Termination
- Military end state
- Objectives
- Effects
- Center of gravity
- Decisive points
- Lines of operation and lines of effort

- Direct and indirect approach
- Anticipation
- Operational reach
- Culmination
- Arranging operations
- Forces and functions

Figure IV-7. Elements of Operational Design

operations. Knowing when to terminate military operations and how to preserve achieved advantages is key to attaining the national strategic end state. To plan effectively for termination, the supported JFC must know how the President and SecDef intend to terminate the joint operation and ensure that the conditions in the OE endure. CCMD campaign plans will not normally have termination criteria.

(2) Termination criteria are developed first among the elements of operational design as they enable the development of the military end state and objectives. Commanders and their staffs must think through, in the early stages of planning, the conditions that must exist in order to terminate military operations on terms favorable to the US and its multinational partners. A hasty or ill-defined end to the operation may bring with it the possibility that the adversary will renew hostilities or other actors may interfere, leading to further conflict. Commanders and their staffs must balance the desire for quick victory with termination on truly favorable terms.

(3) Termination criteria should account for a wide variety of operational tasks that the joint force may need to accomplish, to include disengagement, force protection, transition to post-conflict operations, reconstitution, and redeployment.

(4) Military end states are briefed to SecDef as part of the IPR process to ensure the military end states support the termination criteria. Once approved, the criteria may change. It is important for commanders and staffs to keep an eye out for potential changes, as they may result in a modification to the military end state as well as the commander's operational approach. As such, it is essential for the military to keep a dialogue between the civilian national leadership, and the leadership of other agencies and partners involved.

b. **Military End State.** Military end state is the set of required conditions that defines achievement of all military objectives. It normally represents **a point in time and/or circumstances beyond which the President does not require the military instrument of national power as the primary means to achieve remaining national objectives.** As such, the military end state is often closely tied to termination. While it may mirror many of the conditions of the national strategic end state, the military end state typically will be more specific and contain other supporting conditions. These conditions contribute to developing termination criteria, the specified standards approved by the President and/or SecDef that must be met before a joint operation can be concluded. Aside from its obvious association with strategic or operational objectives, clearly defining the military end state promotes unity of effort, facilitates synchronization, and helps clarify (and may reduce) the risk associated with the campaign or operation. Commanders should include the military end state in their planning guidance and commander's intent statement.

c. **Objectives. An objective is clearly defined, decisive, and attainable.** Once the military end state is understood and termination criteria are established, operational design continues with development of strategic and operational military objectives. Joint planning integrates military actions and capabilities with those of other instruments of national power in time, space, and purpose in unified action to achieve the JFC's military objectives, which contribute to strategic national objectives. Objectives and their supporting effects provide the basis for identifying tasks to be accomplished. **In GEF- and JSCP-directed**

campaign plans, objectives rather than an end state, define the path of the command's actions in contributing to national objectives.

(1) Military missions are conducted to achieve objectives and are linked to national objectives. Military objectives are an important consideration in plan development. They specify what must be accomplished and provide the basis for describing desired effects.

(2) A clear and concise end state allows planners to better examine objectives that must be met to attain the desired end state. Objectives describe what must be achieved to reach or attain the end state. These are usually expressed in military, diplomatic, economic, and informational terms and help define and clarify what military planners must do to support the national strategic end state. Objectives developed at the national-strategic and theater-strategic levels are the defined, decisive, and attainable goals toward which all military operations, activities, and investments are directed within the OA.

(3) Achieving operational objectives ties execution of tactical tasks to reaching the military end state.

(4) There are four primary considerations for an objective.

(a) An objective establishes a single desired result (a goal).

(b) An objective should link directly or indirectly to higher level objectives or to the end state.

(c) An objective is specific and unambiguous.

(d) An objective does not infer ways and/or means—it is not written as a task.

d. **Effects.** An effect is a physical and/or behavioral state of a system that results from an action, a set of actions, or another effect. A desired effect can also be thought of as a condition that can support achieving an associated objective, while an undesired effect is a condition that can inhibit progress toward an objective. In seeking unified action, a JFC synchronizes the military with the diplomatic, informational, and economic power of the US to affect the PMESII systems of relevant actors.

(1) The CCDR plans joint operations based on analysis of national strategic objectives and development of theater strategic objectives supported by measurable strategic and operational desired effects and assessment indicators (see Figure IV-8). At the operational level, a subordinate JFC develops supporting plans, which can include objectives supported by measurable operational-level desired effects and assessment indicators. This may increase operational- and tactical-level understanding of the purpose reflected in the higher-level commander's mission and intent. At the same time, commanders consider potential undesired effects and their impact on the tasks assigned to subordinate commands.

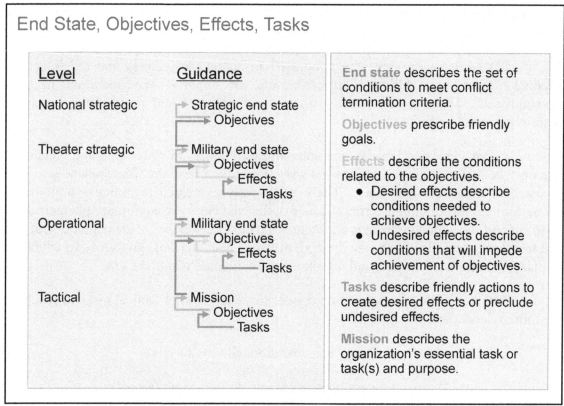

Figure IV-8. End State, Objectives, Effects, Tasks

(2) There are four primary considerations for writing a desired effect statement.

(a) Each desired effect should link directly to one or more objectives.

(b) The effect should be measurable.

(c) The statement should not specify ways and means for accomplishment.

(d) The effect should be distinguishable from the objective it supports as a condition for success, not as another objective or a task.

(3) The proximate cause of effects in complex situations can be difficult to predict particularly when they relate to moral and cognitive issues (such as religion and the "mind of the adversary," respectively). Further, **there will always be gaps in our understanding of the OE.** Commanders and their staffs must appreciate that unpredictable third-party actions, unintended consequences of friendly operations, subordinate initiative and creativity, and the fog and friction of conflict will contribute to an uncertain OE.

(4) The use of effects in planning can help commanders and staff determine the tasks required to achieve objectives and use other elements of operational design more effectively by clarifying the relationships between COGs, LOOs, and/or LOEs; decisive points; and termination criteria. Once a systems perspective of the OE has been developed (and appropriate links and nodes have been identified), the linkage and relationship

between COGs, LOOs, and decisive points can become more obvious. This linkage allows for efficient use of desired effects in planning. The JFC and planners continue to develop and refine desired effects throughout JPP. Monitoring progress toward creating desired effects and avoiding undesired effects continues throughout execution.

(5) A mission is a task or set of tasks, together with the purpose, that clearly indicates the action to be taken and the reason for doing so. It is derived primarily from higher headquarters guidance.

e. **COG**

(1) One of the most important tasks confronting the JFC's staff during planning is identifying and analyzing friendly and adversary COGs. A COG is a source of power that provides moral or physical strength, freedom of action, or will to act. It is what Clausewitz called "the hub of all power and movement, on which everything depends ... the point at which all our energies should be directed." An objective is always linked to a COG. There may also be different COGs at different levels, but they should be nested. At the strategic level, a COG could be a military force, an alliance, political or military leaders, a set of critical capabilities or functions, or national will. At the operational level, a COG often is associated with the adversary's military capabilities—such as a powerful element of the armed forces—but could include other capabilities in the OE. In identifying COGs it is important to remember that irregular warfare focuses on legitimacy and influence over a population, unlike traditional warfare, which employs direct military confrontation to defeat an enemy's armed forces, destroy an enemy's war-making capacity, or seize or retain territory to force a change in an enemy's government or policies. Therefore, during irregular warfare, the enemy and friendly COG may be the same population.

(2) **COGs exist in an adversarial context** involving a clash of moral wills and/or physical strengths.

(a) Since COGs exist only in unitary systems, a CCMD TCP or FCP may not have a single COG, as it may be conducting operations along multiple lines that by themselves are not connected (from the US, they may be looked at as a connected system, but they themselves do not act as a single entity).

(b) COGs are formed out of the relationships between adversaries, and they do not exist in a strategic or operational vacuum. COGs are framed by each party's view of the threats in the OE and the requirements to develop/maintain power and strength relative to their need to be effective in accomplishing their objectives. Therefore, commanders not only must consider the enemy's COGs, but they also must identify and protect their own.

(c) Planners should focus on both the enemy's COGs and the friendly's, and understand that through the conduct of operations COGs may change. Assessment aids in identifying these changes. Because objectives are always linked to COGs, changes in objectives may change the COG for both the adversary and the friendly forces.

(3) The COG construct is useful as an analytical tool to help JFCs and staffs analyze friendly and adversary sources of strength as well as weaknesses and vulnerabilities. This process cannot be taken lightly, since a faulty conclusion resulting from a poor or hasty analysis can have very serious consequences, such as the inability to achieve strategic and operational objectives at an acceptable cost. The selection of COGs is not solely a static process by the J-2 during JIPOE. Planners must continually analyze and refine COGs due to actions taken by friendly forces and the adversary's reactions to those actions. Figure IV-9 shows a number of characteristics that may be associated with a COG.

(4) Analysis of friendly and adversary COGs is a key step in operational design. Joint force intelligence analysts identify adversary COGs, determining from which elements the adversary derives freedom of action, physical strength (means), and the will to fight. The J-2, in conjunction with other operational planners, then attempts to determine if the tentative or candidate COGs truly are critical to the adversary's strategy. This analysis is a linchpin in the planning effort. Others on the joint force staff conduct similar analysis to identify friendly COGs. Once COGs have been identified, JFCs and their staffs determine how to attack enemy COGs while protecting friendly COGs. The protection of friendly strategic COGs such as public opinion and US national capabilities typically requires efforts and capabilities beyond those of just the supported CCDR. An analysis of the identified COGs in terms of critical capabilities, requirements, and vulnerabilities is

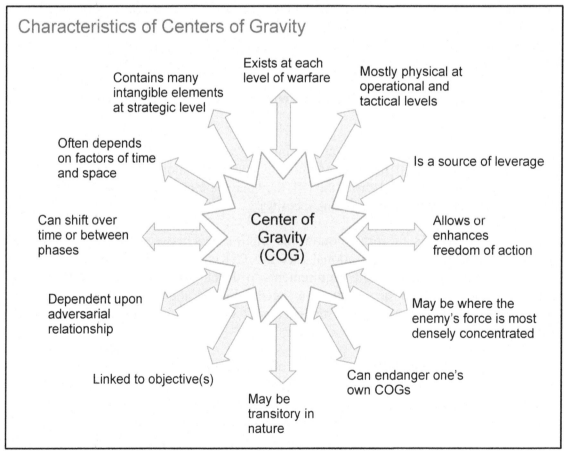

Figure IV-9. Characteristics of Centers of Gravity

vital to this process. As the COG acts as the balance point or focal point that holds the system together, striking it should cause the system to collapse.

(a) Striking the COG and fracturing the system could leave the planner and commander with multiple problems rather than with one, driving up complexity and risk. Even if planners and CCDRs identify a COG, it may not be best to strike it first, if the commander wants to avoid addressing multiple problems or the overall intent is to ensure stability within the system.

(b) Although identifying COGs is useful in understanding the system, planners should refrain from automatically assuming a strike on the COG is the solution to every operation. Consideration must be placed on whether total collapse of the enemy or system is commensurate with the objectives and end state.

(c) Rather, planners may recommend affecting smaller elements of the whole, enabling continued balance until the entire problem is reduced to manageable parts. If the latter approach is taken, planners must take into consideration that as the system changes, the COG, as in the world of physics, may change in relation to the remaining whole.

(d) Similarly, in a cooperative environment such as CCMD campaign plans, operations may be executed to strengthen the COG.

(5) Understanding the relationship among COGs not only permits but also compels greater precision in thought and expression in operational design. Planners should analyze COGs within a framework of three critical factors—capabilities, requirements, and vulnerabilities—to aid in this understanding. **Critical capabilities** are the primary abilities essential to the accomplishment of the objective. **Critical requirements** are essential conditions, resources, and means the COG requires to perform the critical capability. **Critical vulnerabilities** are those aspects or components of critical requirements that are deficient or vulnerable to direct or indirect attack in a manner achieving decisive or significant results. In general, a JFC must possess sufficient operational reach and combat power or other relevant capabilities to take advantage of an adversary's critical vulnerabilities while protecting friendly critical capabilities within the operational reach of an adversary.

(6) When identifying friendly and enemy critical vulnerabilities, the JFC and staff will understandably want to focus their efforts against the critical vulnerabilities that will do the most decisive damage to an enemy's COG. However, in selecting those critical vulnerabilities, planners must also compare their criticality with their accessibility, vulnerability, redundancy, resiliency, and impact on the civilian populace, and then balance those factors against friendly capabilities to affect those vulnerabilities. The JFC should seek opportunities aggressively to apply force against an adversary in as vulnerable an aspect as possible, and in as many dimensions as possible. In other words, the JFC seeks to undermine the adversary's strength by exploiting adversary vulnerabilities while protecting friendly vulnerabilities from adversaries attempting to do the same.

(7) A proper analysis of adversary critical factors must be based on the best available knowledge of how adversaries organize, fight, think, and make decisions, and their physical and psychological strengths and weaknesses. JFCs and their staffs must develop an understanding of their adversaries' capabilities and vulnerabilities, as well as factors that might influence an adversary to abandon its strategic objectives. They must also envision how friendly forces and actions appear from the adversaries' viewpoints. Otherwise, the JFC and the staff may fall into the trap of ascribing to an adversary attitudes, values, and reactions that mirror their own.

(8) Before solidifying COGs into the plan, planners should analyze and test the validity of the COGs. The defeat, destruction, neutralization, or substantial weakening of a valid COG should cause an adversary to change its COA or prevent an adversary from achieving its strategic objectives. If analysis and/or wargaming show this does not occur, then perhaps planners have misidentified the COG, and they must revise their COG and critical factors analysis. The conclusions, while critically important to the planning process itself, must be tempered with continuous evaluations because derived COGs and critical vulnerabilities are subject to change at any time during the campaign or operation. Accordingly, JFCs and their subordinates should be alert to circumstances during execution that may cause derived COGs and critical vulnerabilities to change and adjust friendly plans and operations accordingly.

(9) Commanders must also analyze friendly COGs and identify critical vulnerabilities (see Figure IV-10). For example, long sea and air lines of communications (LOCs) from the continental United States (CONUS) or supporting theaters could be a critical vulnerability for a friendly COG. Through prior planning and coordination, commanders can mitigate the potential impact of challenges such as the failure of foreign governments to provide overflight clearances to US forces or MNFs. A friendly COG could also be something more intangible in nature. During the 1990-1991 Persian Gulf Conflict, for example, the Commander, US Central Command, identified the coalition itself as a friendly operational COG and took appropriate measures to protect it, to include deployment of theater missile defense systems. In conducting the analysis of friendly vulnerabilities, the supported commander must decide how, when, where, and why friendly military forces are (or might become) vulnerable to hostile actions and then plan accordingly. The supported commander must achieve a balance between prosecuting the main effort and protecting critical capabilities and vulnerabilities in the OA to protect friendly COGs.

For more information on COGs and the systems perspective, see JP 2-01.3, Joint Intelligence Preparation of the Operational Environment.

f. **Decisive Points**

(1) A decisive point is a geographic place, specific key event, critical factor, or function that, when acted upon, allows a commander to gain a marked advantage over an enemy or contributes materially to achieving success (e.g., creating a desired effect, achieving an objective). Decisive points can greatly influence the outcome of an action. Decisive points can be physical in nature, such as a constricted sea lane, a hill, a town,

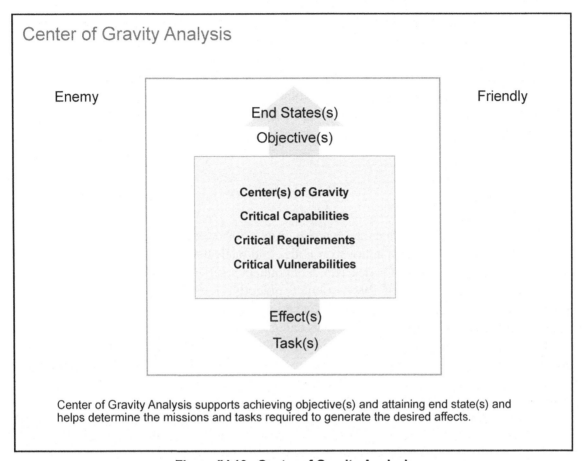

Figure IV-10. Center of Gravity Analysis

weapons of mass destruction material cache or facility, or an air base, but they could include other elements such as command posts, critical boundaries, airspace, or communications and/or intelligence nodes. In some cases, specific key events may also be decisive points, such as attainment of air or maritime superiority, commitment of the enemy's reserve, opening a supply route during humanitarian operations, or gaining the trust of a key leader. In still other cases, decisive points may have a larger systemic impact and, when acted on, can substantially affect the enemy's information, financial, economic, or social systems. When dealing with an irregular threat, commanders and their staffs should consider how actions against decisive points will affect not only the enemy, but also the relevant population's perception of enemy and friendly forces. Collateral effects on the local populace may impact stability in the area or region of interest.

(2) The most important decisive points can be determined from analysis of critical factors. Understanding the relationship between a COG's critical capabilities, requirements, and vulnerabilities can illuminate direct and indirect approaches to the COG. It is likely most of these critical factors will be decisive points, which should then be further addressed in the planning process.

(3) There may often be cases where the JFC's combat power and other capabilities will be insufficient to affect the enemy's COGs rapidly with a single action. In this situation, the supported JFC must selectively focus a series of actions against the

enemy's critical vulnerabilities until the cumulative effects of these actions lead to mission success. Just as a combined arms approach is often the best way to attack an enemy field force in the military system, attacking several vulnerable points in other systems may offer an effective method to influence an enemy COG. The indirect approach may offer the most effective method to exploit enemy critical vulnerabilities through the ID of decisive points. **Although decisive points are usually not COGs, they are the keys to attacking or protecting them.**

g. **LOO and LOE**

(1) **LOOs**

(a) A LOO defines the interior or exterior orientation of the force in relation to the enemy or that connects actions on nodes and/or decisive points related in time and space to an objective(s). LOOs describe and connect a series of decisive actions that lead to control of a geographic or force-oriented objective (see Figure IV-11). Operations designed using LOOs generally consist of a series of actions executed according to a well-defined sequence, although multiple LOOs can exist at the same time (parallel operations). Combat operations are typically planned using LOOs. These lines tie offensive, defensive, and stability tasks to the geographic and positional references in the OA. Commanders synchronize activities along complementary LOOs to attain the military end state.

(b) **A force operates on interior lines when its operations diverge from a central point.** Interior lines usually represent central position, where a friendly force can reinforce or concentrate its elements faster than the enemy force can reposition. With interior lines, friendly forces are closer to separate enemy forces than the enemy forces are to one another. Interior lines allow an isolated force to mass combat power against a specific portion of an enemy force by shifting capabilities more rapidly than the enemy can react.

(c) **A force operates on exterior lines when its operations converge on the enemy.** Operations on exterior lines offer opportunities to encircle and annihilate an enemy force. However, these operations typically require a force stronger or more mobile than the enemy.

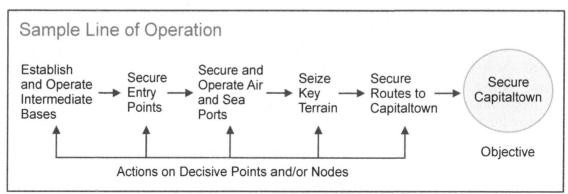

Figure IV-11. Sample Line of Operation

(d) The relevance of interior and exterior lines depends on the time and space relationship between the opposing forces. Although an enemy force may have interior lines with respect to the friendly force, this advantage disappears if the friendly force is more agile and operates at a higher tempo. Conversely, if a smaller friendly force maneuvers to a position between larger but less agile enemy forces, the friendly force may be able to defeat them in detail before they can react effectively.

(2) **LOEs**

(a) A LOE links multiple tasks and missions using the logic of purpose—cause and effect—to focus efforts toward establishing operational and strategic conditions. LOEs are essential to operational design when positional references to an enemy or adversary have little relevance, such as in counterinsurgency operations or stability activities. In operations involving many nonmilitary factors, LOEs may be the only way to link tasks, effects, conditions, and the desired end state (see Figure IV-12). LOEs are often essential to helping commanders visualize how military capabilities can support the other instruments of national power. They are a particularly valuable tool when used to achieve unity of effort in operations involving MNFs and civilian organizations, where unity of command is elusive, if not impractical.

(b) Commanders at all levels may use LOEs to develop missions and tasks and to determine force capability requirements. Commanders synchronize and sequence related actions along multiple LOEs. Seeing these relationships helps commanders assess progress toward attaining the end state as forces perform tasks and accomplish missions.

(c) Commanders typically visualize stability activities along LOEs. For stability activities, commanders may consider linking primary stability tasks to their corresponding DOS post-conflict technical sectors. These stability tasks link military actions with the broader interagency effort across the levels of warfare. A full array of LOEs might include offensive and defensive lines, as well as a line for public affairs, information-related capabilities (IRCs), and counter threat finance. All typically produce effects across multiple LOEs.

(d) Commanders and staff should consider cross-cutting LOEs involving more than one instrument of national power in order to create a more effective system for interagency coordination during execution. LOEs planned around functional areas such as diplomacy or economics create unintentional interagency coordination stovepipes during execution, because they are fixed toward the efforts of a single USG department or agency. Cross-cutting LOEs such as establishing essential services or civil security operations create a tendency toward more dynamic and open interagency coordination during execution because they require the synchronization of efforts of multiple USG departments and agencies. This type of construct brings to bear the capabilities and expertise of multiple elements of the USG, which makes it particularly effective toward achieving more complex objectives.

(3) **Combining LOOs and LOEs.** Commanders may use both LOOs and LOEs to connect objectives to a central, unifying purpose. LOEs can also link objectives,

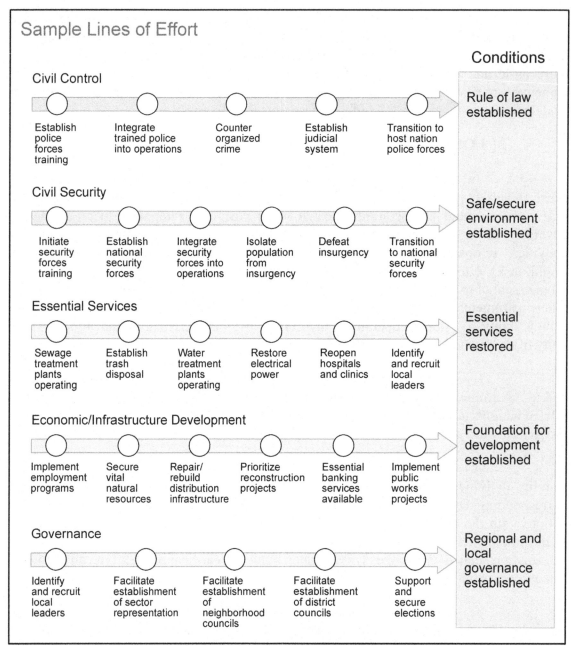

Figure IV-12. Sample Lines of Effort

decisive points, and COGs. Combining LOOs and LOEs allows commanders to include nonmilitary activities in their operational design. This combination helps commanders incorporate stability tasks into their operational approach that are necessary to reach the end state. It allows commanders to consider the less tangible aspects of the OE where the other instruments of national power or nontraditional military activities may dominate. Commanders can then visualize concurrent and post-conflict stability activities. Making these connections relates the tasks, effects, and objectives identified in the operation or campaign plan.

(4) Defeat and stability mechanisms complement COG analysis. While COG analysis helps us understand a problem, defeat and stability mechanisms suggest means to solve it. They provide a useful tool for describing the main effects a commander wants to create along a LOO or LOE.

(a) **Defeat Mechanisms.** Defeat mechanisms primarily apply in combat operations against an active enemy force. Combat aims at defeating armed enemies—regular, irregular, or both, through the organized application of force to kill, destroy, or capture by all means available. There are two basic defeat mechanisms to accomplish this: attrition and disruption. The aim of disruption is to defeat an enemy's ability to fight as a cohesive and coordinated organization. The alternative is to destroy his material capabilities through attrition, which generally is more costly and time-consuming. Although acknowledging that all successful combat involves both mechanisms, joint doctrine conditionally favors disruption because it tends to be a more effective and efficient way of causing an enemy's defeat, and the increasing imperative for restraint in the application of violence may often preclude the alternative. The defeat mechanisms may include:

1. **Destroy.** To identify the most effective way to eliminate enemy capabilities; it may be attained by sequentially applying combat power over time or with a single, decisive attack.

2. **Dislocate.** To compel the enemy to expose forces by reacting to a specific action; it requires enemy commanders to either accept neutralization of part of their force or risk its destruction while repositioning.

3. **Disintegrate.** To exploit the effects of dislocation and destruction to shatter the enemy's coherence; it typically follows destruction and dislocation, coupled with the loss of capabilities that enemy commanders use to develop and maintain situational understanding.

4. **Isolate.** To limit the enemy's ability to conduct operations effectively by marginalizing critical capabilities or limiting the enemy's ability to influence events; it exposes the enemy to continued degradation through the massed effects of other defeat mechanisms.

(b) **Stability Mechanisms.** A stability mechanism is the primary method through which friendly forces affect civilians in order to attain conditions that support establishing a lasting, stable peace. Combinations of stability mechanisms produce complementary and reinforcing effects that help to shape the human dimension of the OE more effectively and efficiently than a single mechanism applied in isolation. Stability mechanisms may include compel, control, influence, and support. Proper application of these stability mechanisms is key in irregular warfare where success is dependent on enabling a local partner to maintain or establish legitimacy and influence over relevant populations.

<u>1.</u> **Compel.** To maintain the threat—or actual use—of lethal or nonlethal force to establish control and dominance; effect behavioral change; or enforce cessation of hostilities, peace agreements, or other arrangements. Legitimacy and compliance are interrelated. While legitimacy is vital to achieving HN compliance, compliance depends on how the local populace perceives the force's ability to exercise force to accomplish the mission. The appropriate and discriminate use of force often forms a central component to success in stability activities; it closely ties to legitimacy. Depending on the circumstances, the threat or use of force can reinforce or complement efforts to stabilize a situation, gain consent, and ensure compliance with mandates and agreements. The misuse of force—or even the perceived threat of the misuse of force—can adversely affect the legitimacy of the mission or the military instrument of national power.

<u>2.</u> **Control.** To establish public order and safety; secure borders, routes, sensitive sites, population centers, and individuals; and physically occupy key terrain and facilities. As a stability mechanism, control closely relates to the primary stability task, establish civil control. However, control is also fundamental to effective, enduring security. When combined with the stability mechanism compel, it is inherent to the activities that comprise disarmament, demobilization, and reintegration, as well as broader security sector reform programs. Without effective control, efforts to establish civil order—including efforts to establish both civil security and control over an area and its population—will not succeed. Establishing control requires time; patience; and coordinated, cooperative efforts across the OA.

<u>3.</u> **Influence.** To alter the opinions and attitudes of the HN population through IRCs, presence, and conduct. It applies nonlethal capabilities to complement and reinforce the compelling and controlling effects of stability mechanisms. Influence aims to effect behavioral change through nonlethal means. It is more a result of public perception than a measure of operational success. It reflects the ability of forces to operate successfully among the people of the HN, interacting with them consistently and positively while accomplishing the mission. Here, consistency of actions, words, and deeds is vital. Influence requires legitimacy. Military forces earn the trust and confidence of the people through the constructive capabilities inherent to combat power, not through lethal or coercive means. Positive influence is absolutely necessary to achieve lasting control and compliance. It contributes to success across the LOEs and engenders support among the people. Once attained, influence is best maintained by consistently exhibiting respect for, and operating within, the cultural and societal norms of the local populace.

<u>4.</u> **Support.** To establish, reinforce, or set the conditions necessary for the other instruments of national power to function effectively, coordinating and cooperating closely with HN civilian agencies and assisting aid organizations as necessary to secure humanitarian access to vulnerable populations. Support is vital to a comprehensive approach to stability activities. The military instrument of national power brings unique expeditionary capabilities to stabilization efforts. These capabilities enable the force to quickly address the immediate needs of the HN and local populace. In extreme circumstances, support may require committing considerable resources for a protracted period. However, easing the burden of support on military forces requires enabling civilian

agencies and organizations to fulfill their respective roles. This is typically achieved by combining the effects of the stability mechanisms—compel, control, and influence—to reestablish security and control, restoring essential civil services to the local populace, and helping to secure humanitarian access necessary for aid organizations to function effectively.

h. **Direct and Indirect Approach.** The approach is the manner in which a commander contends with a COG. A direct approach attacks the enemy's COG or principal strength by applying combat power directly against it. However, COGs are generally well protected and not vulnerable to a direct approach. Thus, commanders usually choose an indirect approach. An indirect approach attacks the enemy's COG by applying combat power against critical vulnerabilities that lead to the defeat of the COG while avoiding enemy strength.

(1) Direct attacks against enemy COGs resulting in their neutralization or destruction provide the most direct path to victory. Since direct attacks against enemy COGs mean attacking an opponent's strength, JFCs must determine if friendly forces possess the power to attack with acceptable risk. Commanders normally attack COGs directly when they have superior forces, a qualitative advantage in leadership, and/or technological superiority over enemy weapon systems. In the event a direct attack is not a reasonable solution, JFCs should consider an indirect approach until conditions are established that permit successful direct attacks (see Figure IV-13). Whenever applicable, JFCs should consider developing simultaneous and/or synchronized action with both direct and indirect approaches. In this manner, the adversary's derived vulnerabilities can offer indirect pathways to gain leverage over its COGs.

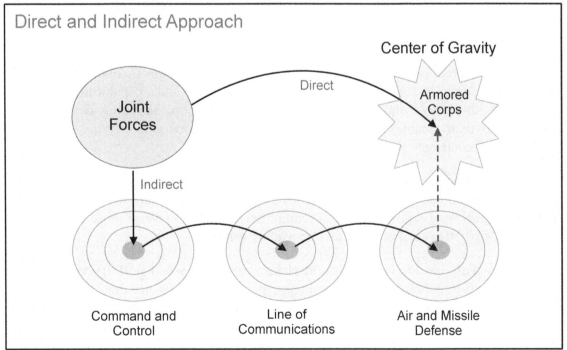

Figure IV-13. Direct and Indirect Approach

(2) At the strategic level, indirect methods of defeating the adversary's COG could include depriving the adversary of allies or friends, emplacing sanctions, weakening the national will to fight by undermining the public support for war, and breaking up cohesion of adversary alliances or coalitions.

(3) At the operational level, the most common indirect method of defeating an enemy's COGs is to conduct a series of attacks against selected aspects of the enemy's combat power. For example, the JFC may sequence combat actions to force an enemy to divide its forces in theater, destroy the enemy's reserves or elements of the enemy's base of operations, or prevent or hinder the deployment of the enemy's major forces or reinforcements into the OA. Indirect methods of attacking the enemy's COGs (through critical vulnerabilities) could entail reducing the enemy's operational reach, isolating the force from its command and control (C2), and destroying or suppressing key protection functions such as air defense. Additionally, in irregular warfare, a persistent indirect approach will help enable a legitimate and capable local partner to address the conflict's causes and to provide security, good governance, and economic development.

i. **Anticipation**

(1) Anticipation is key to effective planning. JFCs must consider what might happen and look for the signs that may bring the possible event to pass. During execution, JFCs should remain alert for the unexpected and for opportunities to exploit the situation. They continually gather information by personally observing and communicating with higher headquarters, subordinates, partner nations, and other organizations in the OA. JFCs may avoid surprise by gaining and maintaining the initiative at all levels of command and throughout the OA, thus forcing the adversary to react rather than initiate, and by thoroughly and continuously wargaming to identify probable adversary reactions to joint force actions. JFCs should also realize the effects of operations and associated consequences on the adversary, interagency and multinational partners, and civilians and prepare for their results.

(2) Shared, common understanding of the OE aids commanders and their staffs in anticipating opportunities and challenges. Knowledge of friendly capabilities; adversary capabilities, intentions, and likely COAs; and the location, activities, and status of dislocated civilians enables commanders to focus joint efforts where they can best, and most directly, contribute to achieving military objectives.

(3) Anticipation is critical to the decision-making process. Operations often require that decisions are made in the advance of need. Decisions such as to mobilize reserves or deploy or reposition forces require anticipation to ensure those requirements are available when needed or when an opportunity arises.

(4) **Anticipation is not without risk.** Commanders and staff officers who tend to lean forward in anticipation of what they expect to encounter are more susceptible to deception efforts by an opponent or having forces out of position if opportunities or threats appear in other places. Therefore, commanders and their staffs should carefully consider all available information upon which decisions are being based. Where possible, multiple

or redundant sources of information should be employed to reduce risk in the decision-making process.

j. **Operational Reach**

(1) Operational reach is the distance and duration across which a joint force can successfully employ military capabilities. Reach may be constrained by the geography, threats, and environmental conditions in and around the OA. Reach may be extended through forward positioning of capabilities and resources, using IRCs, increasing the range and effectiveness of weapon systems, leveraging HNS and contracted support (e.g., system support, external support, theater support) and maximizing the throughput efficiency of the distribution architecture. Operational reach can be unintended. Joint force messages and images may reach outside the OA to unintended audiences creating effects that are contrary to the JFC's objectives. This type of operational reach can be mitigated with properly synchronized communication and proper execution of operations security procedures.

(2) The concept of operational reach is inextricably tied to the concept of LOOs. The geography surrounding and separating our adversaries influences operational reach. Locating forces, reserves, bases, pre-positioned equipment sets, and logistics forward extends operational reach. Operational reach is also affected by increasing the range of weapons and by improving transportation availability and the effectiveness of LOCs and throughput capability. Given the appropriate level of superiority, some assets—such as air, space, and cyberspace—maintain a responsive global capability that significantly extends operational reach. Nevertheless, for any given campaign or major operation, there is a finite range beyond which predominant elements of the joint force cannot prudently operate or maintain effective operations.

(3) Basing, in the broadest sense, is an indispensable part of operational art, since it is tied to the concept of LOOs and directly affects operational reach. Whether from overseas locations, sea-based platforms, or CONUS, basing directly affects the combat power and other capabilities that a joint force can generate. In particular, the arrangement and positioning of advanced bases (often in austere, rapidly emplaced configurations) underwrites the ability of the joint force to shield its components from adversary action and deliver symmetric and asymmetric blows. It also directly influences the combat power and other capabilities the joint force can generate because of its impact on such critical factors as sortie or resupply rates. Political and diplomatic considerations can often affect basing decisions.

(4) US force basing options span the range from permanently based forces to temporary seabasing during crisis response in littoral areas of instability. Bases are typically selected to be within operational reach of the adversary. To that end, analysis during planning must determine whether sufficient infrastructure, destination port and airfield capacities, and diplomatic support exist or can be obtained to support the operational and sustainment requirements of deployed forces, and where they can be assured of some degree of security from attack. Determining where to locate infrastructure and bases poses critical challenges for planners since infrastructure and basing play a key role in enabling campaigns and operations. Adversaries will likely try to develop

antiaccess or area denial capabilities to prevent the buildup and sustainment of forces. One such approach could be a preemptive attack against US forces located outside the adversary's boundaries, so planners must also consider the risk of placing US combat capabilities within the adversary's operational reach. Planners must determine how to mitigate an adversary's efforts to deny access to the theater and its infrastructure and conduct operations as part of the theater campaign to set conditions for future operations.

k. Culmination

(1) Culmination is that point in time and/or space at which the operation can no longer maintain momentum. In the offense, the culminating point is the point at which effectively continuing the attack is no longer possible and the force must consider reverting to a defensive posture or attempting an operational pause. Here the attacker greatly risks counterattack and defeat and continues the attack only at great peril. Success in the attack at all levels is to secure the objective before reaching culmination. A defender reaches culmination when the defending force no longer has the capability to go on the counteroffensive or defend successfully. Success in the defense is to draw the attacker to offensive culmination, then conduct an offensive to expedite the enemy's defensive culmination. During stabilization efforts, culmination may result from the erosion of national will, decline of popular support, questions concerning legitimacy or restraint, or lapses in protection leading to excessive casualties.

(2) The JFC must ensure forces and assets arrive at the right times and places to support the campaign and that sufficient resources will be available when needed in the later stages of the campaign. Integration and synchronization of sustainment with combat operations can forestall culmination and help commanders control the tempo of their operations. At both tactical and operational levels, theater logistic planners forecast the drain on resources associated with conducting operations over extended distance and time. They respond by generating enough military resources at the right times and places to enable their commanders to achieve military strategic and operational objectives before reaching their culminating points. If commanders cannot generate these resources, they should revise their CONOPS.

l. Arranging Operations

(1) Commanders must determine the best arrangement of joint force and component operations to conduct the assigned tasks and joint force mission. This arrangement often will be a combination of simultaneous and sequential operations to reach the end state conditions with the least cost in personnel and other resources. Commanders consider a variety of factors when determining this arrangement, including geography of the OA, available strategic lift, changes in command structure, force protection, distribution and sustainment capabilities, adversary reinforcement capabilities, and public opinion. Thinking about the best arrangement helps determine the tempo of activities in time, space, and purpose. Planners should consider factors such as simultaneity, depth, timing, and tempo when arranging operations.

(a) **Simultaneity refers to the simultaneous application of integrated military and nonmilitary power against the enemy's key capabilities and sources of strength.** Simultaneity in joint force operations contributes directly to an enemy's collapse by placing more demands on enemy forces and functions than can be handled. This does not mean all elements of the joint force are employed with equal priority or that even all elements of the joint force will be employed. It refers specifically to the concept of attacking appropriate enemy forces and functions throughout the OE in such a manner as to damage their morale and physical cohesion.

(b) **Simultaneity also refers to the concurrent conduct of operations at the tactical, operational, and strategic levels.** Tactical commanders fight engagements and battles, understanding their relevance to the contingency plan. JFCs set the conditions for battles within a major operation or campaign to achieve military strategic and operational objectives. GCCs integrate theater strategy and operational art. At the same time, they remain acutely aware of the impact of tactical events. Because of the inherent interrelationships between the various levels of warfare, commanders cannot be concerned only with events at their respective echelon, so commanders at all levels should understand how their actions contribute to the military end state.

(c) The evolution of warfare and advances in technology have expanded the depth of operations. US joint forces can rapidly maneuver over great distances and strike with precision. Joint force operations should be conducted across the full breadth and depth of the OA, creating competing and simultaneous demands on enemy commanders and resources. **The concept of depth seeks to overwhelm the enemy throughout the OA, creating competing and simultaneous demands on enemy commanders and resources and contributing to the enemy's speedy defeat.** Depth applies to time as well as geography. Operations extended in depth shape future conditions and can disrupt an opponent's decision cycle. Global strike, interdiction, and the integration of IRCs with other capabilities are examples of the applications of depth in joint operations. Operations in depth contribute to protection of the force by destroying enemy potential before its capabilities can be realized or employed.

(d) The joint force should conduct operations at a tempo and point in time that maximizes the effectiveness of friendly capabilities and inhibits the adversary. **With proper timing, JFCs can dominate the action, remain unpredictable, and operate beyond the enemy's ability to react.**

(e) The **tempo** of warfare has increased over time as technological advancements and innovative doctrines have been applied to military requirements. While in many situations JFCs may elect to maintain an operational tempo that stretches the capabilities of both friendly and enemy forces, on other occasions JFCs may elect to conduct operations at a reduced pace. During selected phases of a campaign, JFCs could reduce the pace of operations, frustrating enemy commanders while buying time to build a decisive force or tend to other priorities in the OA such as relief to displaced persons. During other phases, JFCs could conduct high-tempo operations designed specifically to overwhelm enemy defensive capabilities. Assuring strategic mobility preserves the JFC's ability to control tempo by allowing freedom of theater access.

(2) Several tools are available to planners to assist with arranging operations. Phases, branches and sequels, operational pauses, and the development of a notional TPFDD all improve the ability of the planner to arrange, manage, and execute complex operations.

(a) **Phases.** Phasing is a way to view and conduct a complex joint operation in manageable parts. The main purpose of phasing is to integrate and synchronize related activities, thereby enhancing flexibility and unity of effort during execution. Reaching the end state often requires arranging an operation or campaign in several phases. Phases in a contingency plan are sequential, but during execution there will often be some simultaneous and overlapping execution of the activities within the phases. In a campaign, each phase can represent a single major operation; while in a major operation, a phase normally consists of several subordinate operations or a series of related activities. See Section D, "Phasing," for a more detailed discussion.

(b) **Branches and Sequels.** Many plans require adjustment beyond the initial stages of the operation. Consequently, JFCs build flexibility into plans by developing branches and sequels to preserve freedom of action in rapidly changing conditions. They are primarily used for changing deployments or direction of movement and accepting or declining combat.

<u>1.</u> **Branches provide a range of alternatives often built into the basic plan.** Branches add flexibility to plans by anticipating situations that could alter the basic plan. Such situations could be a result of adversary action, availability of friendly capabilities or resources, or even a change in the weather or season within the OA.

<u>2.</u> **Sequels anticipate and plan for subsequent operations based on the possible outcomes of the current operation—victory, defeat, or stalemate.**

<u>3.</u> Once the commander and staff have determined possible branches and sequels as far in advance as practicable, they should determine what or where the **decision points** (not to be confused with decisive points) should be. Such decision points capture in space and/or time decisions a commander must make. To aid the commander, planners develop synchronization matrices as well as a decision support matrix (DSM) to link those decision points with the earliest and latest timing of the decision and the appropriate priority intelligence requirements (PIRs) (things the commander must know about the enemy and the OE to make the decision) and friendly force information requirements (FFIRs) (things the commander must know about friendly forces to make the decision). **Each branch from a decision point requires different actions, and each action demands various follow-up actions, such as sequels or potential sequels.**

(c) **Operational Pause**

<u>1.</u> The supported JFC should aggressively conduct operations to obtain and maintain the initiative. However, there may be certain circumstances when this is not feasible because of logistic constraints or force shortfalls. Therefore, **operational pauses may be required when a major operation may be reaching the end of its sustainability.**

As such, operational pauses can provide a safety valve to avoid potential culmination, while the JFC retains the initiative in other ways. However, if an operational pause is properly executed in relation to one's own culminating point, the enemy will not have sufficient combat power to threaten the joint force or regain the initiative during the pause.

2. **Operational pauses are also useful tools for obtaining the proper synchronization of sustainment and operations.** Normally, operational pauses are planned to regenerate combat power or augment sustainment and forces for the next phase, although this will result in extending the duration of a major operation or campaign. Moreover, **properly planned and sequenced operational pauses ensure the JFC has sufficient forces and assets to accomplish strategic or operational objectives.** However, planners must guard against cutting the margin of sustainment and combat effectiveness too thin. Executing a pause before it is necessary provides for flexibility in the timing of the pause and allows for its early termination under urgent conditions without unduly endangering the future effectiveness of the force.

3. Operational pauses can also be times to support strategic decisions, such as offer opportunities for de-escalation and negotiation.

4. **The primary drawback to operational pauses is that they risk forfeiture of strategic or operational initiative.** It is therefore incumbent upon the JFC to plan on as few operational pauses as possible, if any, and consistent with the CONOPS, to alternate pauses and tempo between components of the force. In this manner, a major portion of the joint force can maintain pressure on the enemy through offensive actions while other components pause. Additionally, operational pauses can provide opportunities for military deception if planned in advance.

(d) Realistic plans, branches, sequels, orders, and an accurate TPFDD are important to enable the proper sequencing of operations. Further, the dynamic nature of modern military operations requires adaptability concerning the arrangement of military capabilities in time, space, and purpose. For example, a rapidly changing enemy situation or other aspects of the OE may cause the commander to alter the planned arrangement of operations even as forces are deploying. Therefore, maintaining overall force visibility, to include both in-transit visibility and asset visibility, are critical to maintaining flexibility. The arrangement that the commander chooses should not foreclose future options.

m. **Forces and Functions**

(1) **Commanders and planners can plan campaigns and operations that focus on defeating either enemy forces, functions, or a combination of both.** Typically, JFCs structure operations to attack both enemy forces and functions concurrently to create the greatest possible friction between friendly and enemy forces and capabilities. These types of operations are especially appropriate when friendly forces enjoy technological and/or numerical superiority over an opponent.

(2) JFCs can focus on destroying and disrupting critical enemy functions such as C2, sustainment, and protection. An attack an enemy's functions normally is intended to

destroy the enemy's balance, thereby creating vulnerabilities to be exploited. The direct effect of destroying or disrupting critical enemy functions can create the indirect effects of uncertainty, confusion, and even panic in enemy leadership and forces and may contribute directly to the collapse of enemy capability and will. When determining whether functional attack should be the principal operational approach, JFCs should evaluate several variables within the context of anticipated events such as time required to cripple the enemy's critical functions, time available to the JFC, the enemy's current actions, and likely responses to such actions.

SECTION D. PHASING

16. Application

a. **A phase can be characterized by the focus that is placed on it.** Phases are distinct in time, space, and/or purpose from one another, but must be planned in support of each other and should represent a natural progression and subdivision of the campaign or operation. Each phase should have a set of starting conditions that define the start of the phase and ending conditions that define the end of the phase. The ending conditions of one phase are the starting conditions for the next phase.

b. Phases are necessarily linked and gain significance in the larger context of the campaign. As such, it is imperative the campaign or operation not be broken down into numerous arbitrary components that may inhibit tempo and lead to a plodding, incremental approach. Since a campaign is required whenever pursuit of a strategic objective is not attainable through a single major operation, the theater operational design includes provision for related phases that may or may not be executed.

c. Activities in phases may overlap. The commander's vision of how a campaign or operation should unfold drives subsequent decisions regarding phasing. Phasing, in turn, assists with synchronizing the CONOPS and aids in organizing the assignment of tasks to subordinate commanders. By arranging operations and activities into phases, the JFC can better integrate capabilities and synchronize subordinate operations in time, space, and purpose. Each phase should represent a natural subdivision of the campaign or operation's intermediate objectives. As such, a phase represents a definitive stage during which a large portion of the forces and joint/multinational capabilities are involved in similar or mutually supporting activities.

d. As a general rule, the phasing of the campaign or operation should be conceived in condition-driven rather than time-driven terms. However, resource availability depends in large part on time-constrained activities and factors—such as sustainment or deployment rates—rather than the events associated with the operation. The challenge for planners, then, is to reconcile the reality of time-oriented deployment of forces and sustainment with the condition-driven phasing of operations.

e. Effective phasing must address how the joint force will avoid reaching a culminating point. If resources are insufficient to sustain the force until attaining the end state, planners should consider phasing the campaign or operation to account for necessary operational pauses between phases. Such phasing enables the reconstitution of the joint force during joint operations, but the JFC must understand this may provide the adversary an opportunity to reconstitute as well. In some cases, sustainment requirements, diplomatic factors, and political factors within the HN may even dictate the purpose of certain phases as well as the sequence of those phases. For example, phases may shift the main effort among Service and functional components to maintain momentum while one component is being reconstituted.

f. Commanders determine the number and purpose phases used during a campaign or operation. The use of the phases provides a way to arrange combat and stability activities. **Within the context of these phases established by a higher-level JFC, subordinate JFCs and component commanders may establish additional phases that fit their CONOPS.** For example, the joint force land component commander (JFLCC) or a subordinate commander might have the following four activities inside a phase to seize the initiative: deploy, forcible entry, defense, and offense. The JFLCC could use the offense phase as a transition to the GCC's dominate phase.

17. Number, Sequence, and Overlap

Working within the phasing construct, the actual phases used will vary (compressed, expanded, or omitted entirely) with the joint campaign or operation and will be determined by the JFC. During planning, the JFC establishes conditions, objectives, or events for transitioning from one phase to another and plans sequels and branches for potential contingencies. Phases are designed to be conducted sequentially, but some activities from a phase may begin in a previous phase and continue into subsequent phases. The JFC adjusts the phases to exploit opportunities presented by the adversary or operational situation or to react to unforeseen conditions. A joint campaign or operation may be conducted in multiple phases simultaneously if the OA has widely varying conditions. For instance, the commander may transition to stabilization efforts in some areas while still conducting combat operations in other areas where the enemy has not yet capitulated. Occasionally, operations may revert to a previous phase in an area where a resurgent or new enemy reengages friendly forces.

18. Transitions

Transitions between phases are planned as distinct shifts in focus by the joint force, often accompanied by changes in command or support relationships. The activities that predominate during a given phase, however, rarely align with neatly definable breakpoints. The need to move into another phase is normally identified by assessing that a set of objectives are achieved or that the enemy has acted in a manner that requires a major change in focus for the joint force and is therefore usually event driven, not time driven. Changing the focus of the operation takes time and may require changing commander's objectives, desired effects, measures of effectiveness (MOEs), measures of performance (MOPs), priorities, command relationships, force allocation, or even the approach. An example is the shift of focus from sustained combat operations to a preponderance of stability activities. Hostilities gradually lessen as the joint force facilitates reestablishing order, commerce, and local government and deters adversaries from resuming hostile actions while the US and international community take steps to establish or restore the conditions necessary for long-term stability. This challenge demands an agile shift in joint force skill sets, actions, organizational behaviors, and mental outlooks, and interorganizational coordination with a wider range of interagency and multinational partners and other participants to provide the capabilities necessary to address the mission-specific factors.

CHAPTER V
JOINT PLANNING PROCESS

"In forming the plan of a campaign, it is requisite to foresee everything the enemy may do, and be prepared with the necessary means to counteract it. Plans of the campaign may be modified ad infinitum according to the circumstances, the genius of the general, the character of the troops, and the features of the country."

Napoleon, *Maxims of War*, 1831

1. Introduction

a. JPP is an orderly, analytical set of logical steps to frame a problem; examine a mission; develop, analyze, and compare alternative COAs; select the best COA; and produce a plan or order. The application of operational design provides the conceptual basis for structuring campaigns and operations. JPP provides a proven process to organize the work of the commander, staff, subordinate commanders, and other partners, to develop plans that will appropriately address the problem. It focuses on defining the military mission and development and synchronization of detailed plans to accomplish that mission. Commanders and staffs can apply the thinking methodology introduced in Chapter IV, "Operational Art and Operational Design," to discern the correct mission, develop creative and adaptive CONOPS to accomplish the mission, and synchronize those CONOPS so they can be executed. It applies to both supported and supporting JFCs and to joint force component commands when the components participate in joint planning. Together with operational design, JPP facilitates interaction between and among the commander, staff, and subordinate and supporting headquarters throughout planning. JPP helps commanders and their staffs organize their planning activities, share a common understanding of the mission and commander's intent, and develop effective plans and orders. Figure V-1 shows the primary steps of JPP. The JPP seven-step process aligns with the four APEX planning functions. The first two JPP steps (planning initiation and mission analysis) take place during the APEX strategic guidance planning function. The next four JPP steps (COA development, COA analysis and wargaming, COA comparison, and COA approval) align under the APEX concept development planning function. The final JPP step (plan or order development) occurs during the APEX plan development planning function. While there is no JPP step associated with the APEX plan assessment planning function, plans and orders are assessed with the RATE methodology in mind.

b. JPP is applicable for all planning. Like operational design, it is a logical process to approach a problem and determine a solution. **It is a tool to be used by planners but is not prescriptive.** Based on the nature of the problem, other tools available to the planner, expertise in the planning team, time, and other considerations, the process can be modified as required. Similarly, some JPP steps or tasks may be performed concurrently, truncated, or modified as necessary dependent upon the situation, subject, or time constraints of the planning effort. For example, force planning, as an element of plan development, is different for campaign planning and contingency planning. See subparagraph 9c(3)(b).

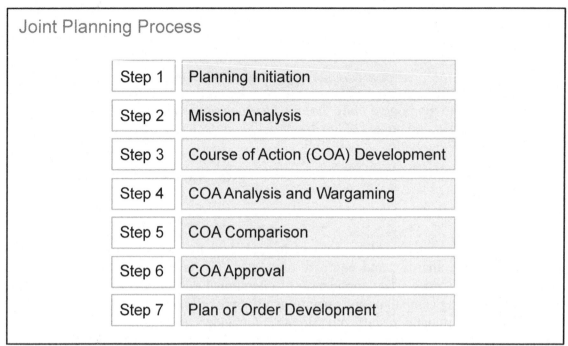

Figure V-1. Joint Planning Process

c. In a crisis, the steps of JPP may be conducted simultaneously to speed the process. Supporting commands and organizations often conduct JPP simultaneously and iteratively with the supported CCMD. In these cases, once mission analysis begins it continues until the operation is complete. Moreover, steps 4-7 are repeated as often as necessary to integrate new requirements (missions) into the development of the plan. This process is depicted in Figure V-2.

2. Operational Art and Operational Design Interface with the Joint Planning Process

a. Operational design and JPP are complementary tools of the overall planning process. Operational design provides an iterative process that allows for the commander's vision and mastery of operational art to help planners answer ends—ways—means—risk questions and appropriately structure campaigns and operations in a dynamic OE. The commander, supported by the staff, gains an understanding of the OE, defines the problem, and develops an operational approach for the campaign or operation through the application of operational design during the initiation step of JPP. Commanders communicate their operational approach to their staff, subordinates, supporting commands, agencies, and multinational/nongovernmental entities as required in their initial planning guidance so that their approach can be translated into executable plans. As JPP is applied, commanders may receive updated guidance, learn more about the OE and the problem, and refine their operational approach. Commanders provide their updated approach to the staff to guide detailed planning. This iterative process facilitates the continuing development

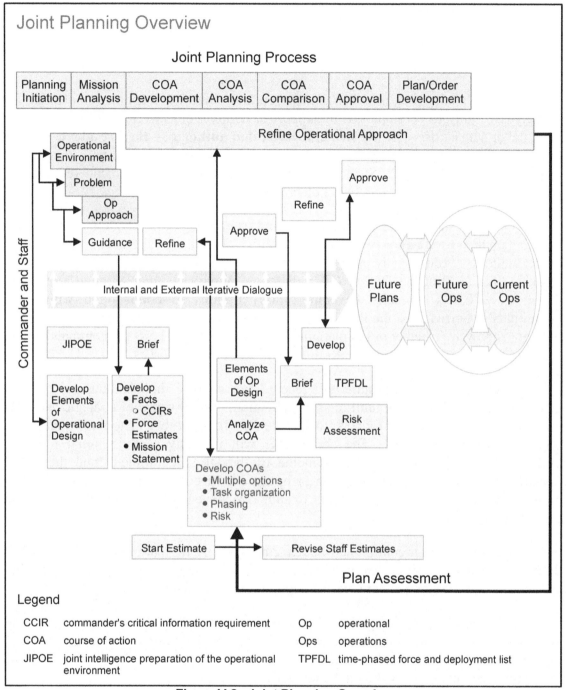

Figure V-2. Joint Planning Overview

and refinement of possible COAs into a selected COA with an associated initial CONOPS and eventually into a resource informed executable plan or order.

b. The relationship between the application of operational art, operational design, and JPP continues throughout the planning and execution of the plan or order. By applying the operational design methodology in combination with the procedural rigor of JPP, the command can monitor the dynamics of the mission and OE while executing operations in accordance with the current approach and revising plans as needed. By combining these approaches, the friendly force can maintain the greatest possible flexibility and do so in a proactive vice reactive manner (Figure V-2).

3. **Planning Initiation (Step 1)**

a. Joint planning begins when an appropriate authority recognizes potential for military capability to be employed in support of national objectives or in response to a potential or actual crisis. **At the strategic level, that authority—the President, SecDef, or CJCS—initiates planning by deciding to develop military options.** Presidential directives, NSS, UCP, GEF, JSCP, and related strategic guidance documents (e.g., SGSs) serve as the primary guidance to begin planning.

b. CCDRs, subordinate commanders, and supporting commanders also initiate planning on their own authority when they identify a planning requirement not directed by higher authority. Additionally, analyses of the OE or developing or immediate crises may result in the President, SecDef, or CJCS directing military planning through a planning directive. CCDRs normally develop military options in combination with other nonmilitary options so that the President can respond with all the appropriate instruments of national power. Whether or not planning begins as described here, the commander may act within approved authorities and ROE/RUF in an immediate crisis.

c. The commander and staff will receive and analyze the planning guidance to determine the time available until mission execution; current status of strategic and staff estimates; and intelligence products, to include JIPOE, and other factors relevant to the specific planning situation. The commander will typically provide initial planning guidance based upon current understanding of the OE, the problem, and the initial operational approach for the campaign or operation. It could specify time constraints, outline initial coordination requirements, or authorize movement of key capabilities within the JFC's authority.

d. While planning is continuous once execution begins, it is particularly relevant when there is new strategic direction, significant changes to the current mission or planning assumptions, or the commander receives a mission for follow-on operations.

e. Planning for campaign plans is different from contingency plans in that contingency planning focuses on the anticipation of future events, while campaign planning assesses the current state of the OE and identifies how the command can shape the OE to deter crisis on a daily basis and support strategic objectives.

4. **Mission Analysis (Step 2)**

a. The CCDR and staff analyzes the strategic direction and derives the restated mission statement for the commander's approval, which allows subordinate and supporting commanders to begin their own estimates and planning efforts for higher headquarters' concurrence. **The joint force's mission is the task or set of tasks, together with the purpose, that clearly indicates the action to be taken and the reason for doing so.** Mission analysis is used to study the assigned tasks and to identify all other tasks necessary to accomplish the mission. Mission analysis is critical because it provides direction to the commander and the staff, enabling them to focus effectively on the problem at hand. When the commander receives a mission tasking, analysis begins with the following questions:

(1) What is the purpose of the mission received? (What problem is the commander being asked to solve or what change to the OE is desired?)

(2) What tasks must my command do for the mission to be accomplished?

(3) Will the mission achieve the desired results?

(4) What limitations have been placed on my own forces' actions?

(5) What forces/assets are needed to support my operation?

(6) How will I know when the mission is accomplished successfully?

b. **The primary inputs to mission analysis** are strategic guidance; the higher headquarters' planning directive; and the commander's initial planning guidance, which may include a description of the OE, a definition of the problem, the operational approach, initial intent, and the JIPOE (see Figure V-3). **The primary products of mission analysis** are staff estimates, the mission statement, a refined operational approach, the commander's intent statement, updated planning guidance, and initial CCIRs.

c. Mission analysis helps the JFC understand the problem and purpose of the operation and issue appropriate guidance to drive the rest of the planning process. The JFC and staff can accomplish mission analysis through a number of logical activities, such as those shown in Figure V-4.

(1) Although some activities occur before others, mission analysis typically involves substantial concurrent processing of information by the commander and staff, particularly in a crisis situation.

(2) During mission analysis, it is essential the tasks (specified and implied) and their purposes are clearly stated to ensure planning encompasses all requirements, limitations (constraints—must do, or restraints—cannot do) on actions that the commander or subordinate forces may take are understood, and the correlation between the commander's mission and intent and those of higher and other commanders is understood. Resources and authorities must also be evaluated to ensure there is not a mission-resource-authority mismatch and second, to enable the commander to prioritize missions and tasks against limited resources.

(3) Additionally, during mission analysis, specific information may need to be captured and tracked in order to improve the end products. This includes requests for information regarding forces, capabilities, and other resources; questions for the commander or special assistant (e.g., legal); and proposed battle rhythm for planning and execution. Recording this information during the mission analysis process will enable a more complete product and smoother mission analysis brief.

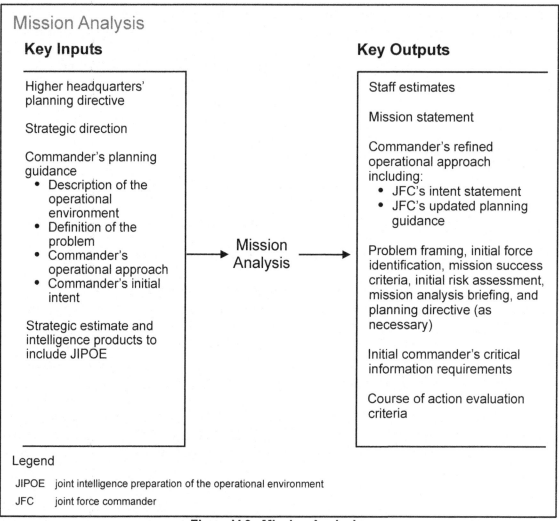

Mission Analysis

Key Inputs

Higher headquarters' planning directive

Strategic direction

Commander's planning guidance
- Description of the operational environment
- Definition of the problem
- Commander's operational approach
- Commander's initial intent

Strategic estimate and intelligence products to include JIPOE

Mission Analysis

Key Outputs

Staff estimates

Mission statement

Commander's refined operational approach including:
- JFC's intent statement
- JFC's updated planning guidance

Problem framing, initial force identification, mission success criteria, initial risk assessment, mission analysis briefing, and planning directive (as necessary)

Initial commander's critical information requirements

Course of action evaluation criteria

Legend

JIPOE   joint intelligence preparation of the operational environment
JFC   joint force commander

Figure V-3. Mission Analysis

d. **Analyze Higher Headquarters' Planning Directives and Strategic Guidance**

(1) Strategic guidance is essential to joint planning and operational design. The President, SecDef, and CJCS promulgate strategic direction documents that cover a broad range of situations, and CCDRs provide guidance that covers a more narrow range of theater or functional situations. Documents such as the UCP, GEF, and JSCP provide near-term (0-2 years) strategic direction, and the CCDR's theater or functional strategy provide the mid- to long-term (greater than 3 years) GCC or FCC vision for the AOR or global employment of functional capabilities prepared in the context of SecDef's priorities. CCDR strategy links national strategic direction to joint planning.

(2) For a specific crisis, an order provides specific guidance, typically including a description of the situation, purpose of military operations, objectives, anticipated mission or tasks, and pertinent limitations. The GFMIG apportionment tables identify forces planners can reasonably expect to be available. Supported and supporting plans for the same military activity are constrained to the same resources. Planners should not expect to use additional forces beyond those listed in the apportionment tables without CJCS

Mission Analysis Activities

- Begin logistics supportability analysis.

- Analyze higher headquarters planning activities and strategic guidance.

- Review commander's initial planning guidance, including his initial understanding of the operational environment, of the problem, and description of the operational approach.

- Determine known facts and develop planning assumptions.

- Determine and analyze operational limitations.

- Determine specified, implied, and essential tasks.

- Develop mission statement.

- Conduct initial force identification.

- Develop risk assessment.

- Develop mission success criteria.

- Develop commander's critical information requirements.

- Prepare staff estimates.

- Prepare and deliver mission analysis brief.

- Publish commander's updated planning guidance, intent statement, and refined operational approach.

Steps are not necessarily sequential.

Figure V-4. Mission Analysis Activities

approval. The CJCS may amplify apportionment guidance for the specific crisis. This planning can confirm or modify the guidance for an existing contingency plan or order. This might simplify the analysis step, since consensus should already exist between the supported command and higher authority on the nature of the OE in the potential joint operations area (JOA)—such as the political, economic, social, and military circumstances—and potential US or multinational responses to various situations described in the existing plan. But even with a preexisting contingency plan, planners need to confirm the actual situation matches the hypothetical situation that the contingency plan was based on, as well as validating other assumptions. Significant changes may require refining or adapting the existing contingency plan. The dynamic nature of an emerging crisis can change many key aspects of the OE compared with earlier assumptions. These changes can greatly affect the plan's original operational approach upon which the commander and staff based decisions about COA alternatives and tasks to potential subordinate and supporting commands. In particular, planners must continuously monitor, assess, and adjust the strategic and operational objectives, planning assumptions, and criteria that comprise the military objectives. Differences between the commander's perspective and that of higher headquarters must be resolved at the earliest opportunity.

---

(3) In time-compressed situations, especially with no preexisting plan, the higher headquarters' assessment of the OE and objectives may be the only guidance available. However, this circumstance is one that can benefit the most from the commander's and staff's independent assessment of circumstances to ensure they share a common understanding with higher headquarters assessment of the OE, strategic objectives, and the tasks or mission assigned to achieve these objectives. This is why CCMD JIPOE efforts should be continuous; these efforts maintain the intelligence portions of the CCDR's strategic estimate. Keeping the strategic estimate up to date greatly facilitates planning in a crisis as well as the transition of contingency plans to execution in crisis situations.

(4) **Multinational Strategic Guidance.** CCDRs, JFCs, component and supporting commanders, and their staffs must clearly understand both US and partner nation strategic and military objectives and conditions that the national or multinational political leadership want the multinational military force to attain in terms of the internal and external balance of power, regional security, and geopolitics. To ensure unity of effort, planners should identify and attempt to resolve conflicts between participating nations' objectives and identify possible conflicts between different nations' national political and military objectives to ensure strategic planning accounts for these divergences. When multinational objectives are unclear, the senior US military commander must seek clarification and convey the positive or negative impact of continued ambiguity to the President and SecDef. For additional information on multinational operations, see JP 3-16, *Multinational Operations.* For specific information on NATO operations, see Allied Joint Publication (AJP)-01, *Allied Joint Doctrine;* AJP-3, *Allied Joint Doctrine for the Conduct of Operations;* and AJP-5, *Allied Joint Doctrine for Operational-Level Planning.*

e. **Review Commander's Initial Planning Guidance.** Staff members and representatives from supporting organizations should maintain an open dialogue with the commander to better develop an appropriate solution to the problem and be able to adapt solutions to match the evolving OE and any potentially changing problems. Staffs should analyze the CCDR's initial planning guidance for the campaign or operation, which provides a basis for continued detailed analysis of the OE and of the tasks that may describe the mission and its parameters.

> Commanders and planners must use caution in characterizing information as facts, as some items of information thought to be facts may be open to interpretation, based on the observer's perspective or incomplete information.

f. **Determine Known Facts and Develop Planning Assumptions.** The staff assembles both facts and assumptions to support the planning process and planning guidance.

(1) A fact is a statement of information known to be true (such as verified locations of friendly and adversary force dispositions).

(2) An assumption provides a supposition about the current situation or future course of events, presumed to be true in the absence of facts. A valid assumption can be

developed for both friendly and adversary situations and has three characteristics: logical, realistic, and essential for planning to continue. Commanders and staffs should never assume away adversary capabilities or assume that unrealistic friendly capabilities would be available. Assumptions address gaps in knowledge that are critical for the planning process to continue. **Assumptions must be continually reviewed to ensure validity and challenged if they appear unrealistic.** Subordinate commanders must not develop assumptions that contradict valid higher headquarters assumptions.

(a) Commanders and staffs should anticipate changes to the plan if an assumption proves to be incorrect. Because of assumptions' influence on planning, planners must either validate the assumptions (treat as facts) or invalidate the assumptions (alter the plan accordingly) as quickly as possible.

(b) During wargaming or red teaming, planners should review both the positive and negative aspect of all assumptions. They should review the plan from both the perspective that the assumption will prove true and from the perspective that the assumption will prove false. This can aid in preventing biases or tunnel vision during crisis action procedures.

For more discussion on red teams, see Appendix K, "Red Teams."

(c) Assumptions made in contingency planning should be addressed in the plan. Activities and operations in the plan can be used to validate, refute, or render unnecessary contingency plan assumptions.

(d) Plans may contain assumptions that cannot be resolved until a potential crisis develops. As a crisis develops, assumptions should be replaced with facts as soon as possible. The staff accomplishes this by identifying the information needed to validate assumptions and submitting an information request to an appropriate agency as an information requirement. If the commander needs the information to make a key decision, the information requirement can be designated a CCIR. Although there may be exceptions, the staff should strive to resolve all assumptions before issuing the OPORD.

(e) Planners should attempt to use as few assumptions as necessary to continue planning. By definition, assumptions introduce possibility for error. If the assumption is not necessary to continue planning, its only effect is to introduce error and add the likelihood of creating a bias in the commander's and planner's perspective. Since most plans require refinement, a simpler plan with fewer assumptions allows the commander and staff to act and react with other elements of the OE (including adversaries, allies, and the physical element). However, assumptions can be useful to identify those issues the commander and planners must validate on execution.

(f) All assumptions should be identified in the plan or decision matrix to ensure they are reviewed and validated prior to execution.

g. **Determine and Analyze Operational Limitations.** Operational limitations are actions required or prohibited by higher authority and other restrictions that limit the commander's freedom of action, such as diplomatic agreements, political and economic conditions in affected countries, and partner nation and HN issues.

(1) A **constraint** is a requirement, "must do," placed on the command by a higher command that **dictates an action,** thus restricting freedom of action. For example, General Eisenhower was required to enter the continent of Europe instead of relying upon strategic bombing to defeat Germany.

(2) A **restraint** is a requirement, "cannot do," placed on the command by a higher command that **prohibits an action,** thus restricting freedom of action. For example, General MacArthur was prohibited from striking Chinese targets north of the Yalu River during the Korean War.

(3) Many operational limitations are commonly expressed as ROE. Operational limitations may restrict or bind COA selection or may even impede implementation of the chosen COA. Commanders must examine the operational limitations imposed on them, understand their impacts, and develop options within these limitations to promote maximum freedom of action during execution.

(4) Other operational limitations may arise from laws or authorities, such as the use of specific types of funds or training events. Commanders are responsible for ensuring they have the authority to execute operations and activities.

h. **Determine Specified, Implied, and Essential Tasks.** The commander and staff will typically review the planning directive's specified tasks and discuss implied tasks during planning initiation to resolve unclear or incorrectly assigned tasks with higher headquarters. If there are no issues, the commander and staff will confirm the tasks in mission analysis and then develop the initial mission statement.

(1) **Specified tasks** are those that a commander assigns to a subordinate commander in a planning directive. These are tasks the commander wants the subordinate commander to accomplish, usually because they are important to the higher command's mission and/or objectives. One or more specified tasks often become essential tasks for the subordinate commander.

EXAMPLES OF SPECIFIED TASKS

Ensure freedom of navigation for US forces through the Strait of Gibraltar.

Defend Country Green against attack from Country Red.

(2) **Implied tasks** are additional tasks the commander must accomplish, typically in order to accomplish the specified and essential tasks, support another command, or otherwise accomplish activities relevant to the operation or achieving the objective. In addition to the higher headquarters' planning directive, the commander and staff will review other sources of guidance for implied tasks, such as multinational planning documents and the GCC's TCP, FCPs, enemy and friendly COG analysis products, JIPOE products, relevant doctrinal publications, interviews with subject matter experts, and the

commander's operational approach. The commander can also deduce implied tasks from knowledge of the OE, such as the enemy situation and political conditions in the assigned OA. However, implied tasks do not include routine tasks or SOPs that are inherent in most operations, such as conducting reconnaissance and protecting a flank.

EXAMPLES OF IMPLIED TASKS

Establish maritime superiority out to 50 miles from the Strait of Gibraltar.

Be prepared to conduct foreign internal defense and security force assistance operations to enhance the capacity and capability of Country Green security forces to provide stability and security if a regime change occurs in Country Red.

---

(3) **Essential tasks** are those that the command must execute successfully to attain the desired end state defined in the planning directive. The commander and staff determine essential tasks from the lists of both specified and implied tasks. Depending on the scope of the operation and its purpose, the commander may synthesize certain specified and implied task statements into an essential task statement. See the example mission statement below for examples of essential tasks.

i. **Develop Mission Statement.** The mission statement describes the mission in terms of the elements of who, what, when, where, and why. The commander's operational approach informs the mission statement and helps form the basis for planning. The commander includes the mission statement in the planning guidance, planning directive, staff estimates, commander's estimate, CONOPS, and completed plan.

EXAMPLE MISSION STATEMENT

When directed [when], United States X Command, in concert with coalition partners [who], deters Country Y from coercing its neighbors and proliferating weapons of mass destruction [what] in order to maintain security [why] in the region [where].

---

j. **Conduct Initial Force and Resource Analysis**

(1) **Initial Force Analysis.** During mission analysis, the planning team begins to develop a rough-order of magnitude list of required forces and capabilities necessary to accomplish the specified and implied tasks. Planners consider the responsiveness of assigned and currently allocated forces. While more deliberate force requirement ID efforts continue during concept and plan development, initial ID of readily available forces during mission analysis may constrain the scope of the proposed operational approach.

(a) Force requirements for a plan are initially documented in a force list developed from forces that are assigned, allocated, and apportioned. The force list may be an informal list (TPFDL) and later in the planning process entered into an information technology system such as JOPES as a baseline of forces to support subsequent time phasing. Planners should consider, at the onset of planning, that plan force requirements should be documented in a format and system that enables GFM allocation should the plan transition to execution.

(b) In a crisis, assigned and allocated forces currently deployed to the geographic CCMD's AOR may be the most responsive during the early stages of an emergent crisis. Planners may consider assigned forces as likely to be available to conduct activities unless allocated to a higher priority. Re-missioning previously allocated forces may require SecDef approval and should be coordinated through the JS using procedures outlined CJCSM 3130.06, *(U) Global Force Management Allocation Policies and Procedures.*

(c) Planners should also identify the status of reserve forces and identify the time required for call up and mobilization.

(d) Planners should evaluate appropriate requirements against existing or potential contracts or task orders to determine if the contracted support solution could meet the requirements.

(e) Planners must take into consideration force requirements for supported and supporting plans are drawing from the same quantity of apportioned forces and will compete with requirements for military activities and ongoing operations when the plan is executed.

(f) Finally, planners compare the specified and implied tasks to the forces and resources available and identify shortfalls.

(2) **Identify Non-Force Resources Available for Planning.** In many types of operations, the commander (and planners) may have access to non-force resources, such as commander's initiative funds, other funding sources (such as train and equip funding, support to foreign security forces funding, etc.), or can work with other security assistance programs (foreign military sales, excess defense article transfers, etc.). Planners and commanders can weave together resources and authorities from several different programs to create successful operations.

See JP 3-20, Security Cooperation, *for additional information on integrating multiple resources. See the GFMIG, for more information on the GFM processes and CJCSM 3130.06,* (U) Global Force Management Allocation Policies and Procedures, *for additional guidance on GFM allocation.*

k. **Develop Mission Success Criteria**

(1) **Mission success criteria** describe the standards for determining mission accomplishment. The JFC includes these criteria in the initial planning guidance so the joint force staff and components better understand what constitutes mission success. Mission success criteria apply to all joint operations. Specific success criteria can be utilized for development of supporting objectives, effects, and tasks and therefore become the basis for operation assessment. These also help the JFC determine if and when to move

to the next phase. The initial set of criteria determined during mission analysis becomes the basis for operation assessment.

(2) If the mission is unambiguous and limited in time and scope, mission success criteria can derive directly from the mission statement. For example, if the JFC's mission is to "evacuate all US personnel from the US Embassy in Grayland," then mission analysis could identify two primary success criteria: all US personnel are evacuated and established ROE are not violated.

(3) However, more complex operations will require more complex assessments with MOEs and MOPs for each task, effect, and phase of the operation. These measures must evaluate not only the success of the specific task or mission, but that the desired objective was achieved (the conditions in the OE are those in support of US objectives or interests).

(4) Campaigns and complex operations will often require multiple phases or steps to accomplish the mission. Planners can use a variety of methods through a developed operational approach to identify progress toward the desired objective or end state. Attainment of objectives is one method to assess progress. Commanders review MOEs and MOPs as two additional methods to measure success.

(5) Measuring the status of tasks, effects, and objectives becomes the basis for reports to senior commanders and civilian leaders on the progress of the operation. The CCDR can then advise the President and SecDef accordingly and adjust operations as required. Whether in a supported or supporting role, JFCs at all levels must develop their mission success criteria with a clear understanding of termination criteria established by the CJCS and SecDef. Commanders and staffs should be aware that successful accomplishment of the task or objective might not produce the desired results—and be ready to make recommendations to the President or SecDef on changes to the campaign or operation.

See Chapter VI, "Operation Assessment," and Appendix D, "Operation Assessment Plan (Examples)," for more information on operation assessments.

l. **Develop COA Evaluation Criteria.** Evaluation criteria are standards the commander and staff will later use to measure the relative effectiveness and efficiency of one COA relative to other COAs. Developing these criteria during mission analysis or as part of commander's planning guidance helps to eliminate a source of bias prior to COA analysis and comparison. Evaluation criteria address factors that affect success and those that can cause failure. Criteria change from mission to mission and must be clearly defined and understood by all staff members before starting the wargame to test the proposed COAs. Normally, the chief of staff (COS) (or executive officer) initially determines each proposed criterion with weights based on its relative importance and the commander's guidance. Commanders adjust criterion selection and weighting according to their own experience and vision. The staff member responsible for a functional area scores each COA using those criteria. The staff presents the proposed evaluation criteria to the commander at the mission analysis brief for approval.

m. **Develop Risk Assessment**

(1) Planners conducting a preliminary risk assessment must identify the obstacles or actions that may preclude mission accomplishment and then assess the impact of these impediments to the mission. Once planners identify the obstacles or actions, they assess the probability of achieving objectives and severity of loss linked to an obstacle or action, and characterize the military risk. Based on judgment, military risk assessment is an integration of probability and consequence of an identified impediment.

(2) The probability of the impediment occurring may be ranked as **very likely:** occurs often, continuously experienced; **likely:** occurs several times; **questionable:** unlikely, but could occur at some time; or **unlikely:** can assume it will not occur. Based on probabilities, military risk (consequence) may be **high:** critical objectives cannot be achieved; **significant:** only the most critical objectives can be achieved; **moderate:** can partially achieve all objectives; or **low:** can fully achieve all objectives.

(3) Determining military risk is more an art than a science. Planners use historical data, intuitive analysis, and judgment. Military risk characterization is based on an evaluation of the probability that the commander's objectives will be accomplished. The level of risk is **high** if achieving objectives or obtaining end states is unlikely, **significant** if achieving objectives or obtaining end states is questionable, **moderate** if achieving objectives or obtaining end states is likely, and **low** if achieving objectives or obtaining end states is very likely.

(4) Planners and commanders need to be able to explain military risk to civilian leadership who may not be as familiar with military operations as they are. Additionally, since military risk is often a matter of perspective and personal experience, they must be able to help decision makers understand how they evaluated the probability of accomplishing objectives, how they characterized the resultant military risk, and the sources or causes of that risk.

(5) During decision briefs, risks must be explained using standard terms that support the decision-making process, such as **mission success** (which missions will and which will not be accomplished), **time** (how much longer will a mission take to achieve success), and **forces** (casualties, future readiness, etc.), and political implications.

n. **Determine Initial CCIRs**

(1) **CCIRs are elements of information the commander identifies as being critical to timely decision making.** CCIRs help focus information management and help the commander assess the OE, validate (or refute) assumptions, identify accomplishment of intermediate objectives, and identify decision points during operations. **CCIRs belong exclusively to the commander. They are situation-dependent, focused on predictable events or activities, time-sensitive, and always established by an order or plan.** The CCIR list is normally short so that the staff can focus its efforts and allocate scarce resources. The CCIR list is not static; JFCs add, delete, adjust, and update CCIRs throughout plan development, assessment, and execution based on the information they

need for decision making. PIRs and FFIRs constitute the total list of CCIRs (see Figure V-5).

(a) **PIRs** focus on the adversary and the OE and are tied to commander's decision points. They drive the collection of information by all elements of a command requests for national-level intelligence support and requirements for additional intelligence capabilities. All staff sections can recommend potential PIRs they believe meet the commander's guidance. However, the joint force J-2 has overall staff responsibility for consolidating PIR nominations and for providing the staff recommendation to the commander. **JFC-approved PIRs are automatically CCIRs.**

For more information on PIRs, see JP 2-0, Joint Intelligence.

(b) **FFIRs** focus on information the JFC must have to assess the status of the friendly force and supporting capabilities. All staff sections can recommend potential FFIRs they believe meet the commander's guidance. Commander-approved FFIRs are automatically CCIRs.

(2) A CCIR must be a decision required of the commander, not of the staff, and responding to a CCIR must be critical to the success of the mission.

(3) **Decision Support.** CCIRs support the commander's future decision requirements and are often related to MOEs and MOPs. PIRs are often expressed in terms of the elements of PMESII while FFIRs are often expressed in terms of the diplomatic,

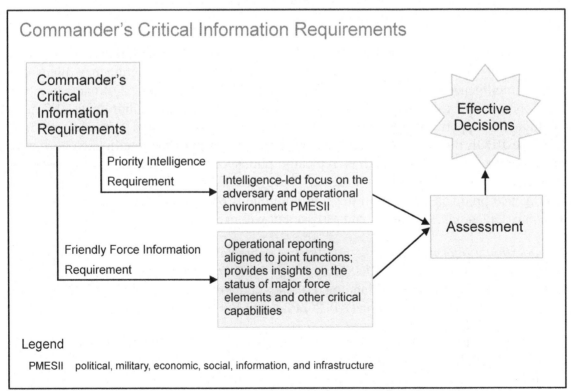

Figure V-5. Commander's Critical Information Requirements

informational, military, and economic instruments of national power. All are developed to support specific decisions the commander must make.

o. **Prepare Staff Estimates**

(1) A staff estimate is an evaluation of how factors in a staff section's functional area support and impact the mission. The purpose of the staff estimate is to inform the commander, staff, and subordinate commands how the functional area supports mission accomplishment and to support COA development and selection.

(2) Staff estimates are initiated during mission analysis, at which point functional planners are focused on collecting information from their functional areas to help the commander and staff understand the situation and conduct mission analysis. Later, during COA development and selection, functional planners fully develop their estimates providing functional analysis of the COAs, as well as recommendations on which COAs are supportable. They should also identify critical shortfalls or obstacles that impact mission accomplishment. Staff estimates are continually updated based on changes in the situation. Operation assessment provides the means to maintain running staff estimates for each functional area.

(3) Not every situation will require or permit a lengthy and formal staff estimate process. In a crisis, staff estimates may be given orally to support the rapid development of plans. However, with sufficient time, planning will demand a more formal and thorough process. Staff estimates should be shared with subordinate and supporting commanders to help them prepare their supporting estimates, plans, and orders. This will improve parallel planning and collaboration efforts of subordinate and supporting elements and help reduce the planning times for the entire process.

(4) Intelligence support to joint planning includes Defense Intelligence Agency-produced dynamic threat assessments (DTAs) for top-priority contingency plans and theater intelligence assessments with a 2-5 year outlook to support CCDR campaign plan development and assessment. Additionally, CCMD JIOCs and subordinate JFC joint intelligence support elements produce intelligence assessments and estimates resulting from the JIPOE process. The intelligence estimate constitutes the intelligence portion of the commander's estimate and is typically published as Appendix 11 to Annex B (Intelligence) to a plan or an order. These are baseline information and finished intelligence products that inform the four continuous operational activities of situational awareness, planning, execution, and assessment within APEX.

For additional information on the intelligence estimate format and its relationship to the commander's estimates, see CJCSM 3122.01, Joint Operation Planning and Execution System (JOPES) Volume I (Planning Policies and Procedures), and CJCSM 3130.03, Adaptive Planning and Execution (APEX) Planning Formats and Guidance.

(5) During mission analysis, intelligence planners lead the development of PIRs to close critical knowledge gaps in initial estimative intelligence products or to validate threat and OE-related planning assumptions. Throughout JPP, additional PIRs may be

nominated to support critical decisions needed throughout all phases of the operation. The intelligence planner then prepares a J-2 staff estimate, which is an appraisal of available capabilities within the intelligence joint function to satisfy commanders' PIRs. This estimate drives development of Annex B (Intelligence) to a plan or an order. In Annex B, the J-2 publishes the commanders PIRs, describes the concept of intelligence operations, specifies intelligence procedures, and assigns intelligence tasks to subordinate and supporting agencies. Tasks assigned to supporting agencies may result in the development of CSA supporting plans through which CSA directors present supporting capabilities to CCDRs for employment in either a deployed or reachback mode. Through the intelligence planning (IP) process, intelligence planners identify gaps and shortfalls in DOD intelligence capabilities. Should these be left unmitigated, they may present risks to the execution of the supported plan to be considered during APEX plan assessment.

For additional information on the IP process, see JP 2-0, Joint Intelligence, *and CJCSM 3314.01,* Intelligence Planning.

(6) The commander's logistics staff and Service component logisticians should develop a logistics overview, which includes but is not restricted to critical logistics facts, assumptions, and information requirements that must be incorporated into the CCIRs; current or anticipated HNS and status; ID of existing contracts and task orders available for use; identifying aerial and sea ports of debarkation; any other distribution infrastructure and associated capacity; inventory (e.g., on-hand, prepositioned, theater reserve); combat support and combat service support capabilities; known or potential capability shortfalls; and contractor support required to replace or augment unavailable military capabilities. From this TLO, a logistics estimate can identify known and anticipated factors that may influence the logistics support.

For more information on estimates, see Appendix C, "Staff Estimates." CJCSM 3122.01, Joint Operation Planning and Execution System (JOPES) Volume I (Planning Policies and Procedures), *contains sample formats for staff estimates.*

p. **Prepare and Deliver Mission Analysis Brief**

(1) Upon conclusion of the mission analysis, the staff will present a mission analysis brief to the commander. This brief provides the commander with the results of the staff's analysis of the mission, offers a forum to discuss issues that have been identified, and ensures the commander and staff share a common understanding of the mission. The results inform the commander's development of the mission statement. The commander provides refined planning guidance and intent to guide subsequent planning. Figure V-6 shows an example mission analysis briefing.

(2) The mission analysis briefing may be the only time the entire staff is present and the only opportunity to make certain all staff members start from a common reference point. The briefing focuses on relevant conclusions reached as a result of the mission analysis.

Example Mission Analysis Briefing

- Introduction
- Situation overview
  - Operational environment (including joint operations area) and threat overview
  - Political, military, economic, social, information, and infrastructure strengths and weaknesses
  - Enemy (including center[s] of gravity) and objectives

- Friendly assessment
  - Facts and assumptions
  - Limitations—constraints/restraints
  - Capabilities allocated
  - Legal considerations

- Communication synchronization
- Objectives, effects, and task analysis
  - United States Government interagency objectives
  - Higher commander's objectives/mission/guidance
  - Objectives and effects
  - Specified/implied/essential tasks
  - Centers of gravity

- Operational protection
  - Operational risk
  - Mitigation

- Proposed initial commander's critical information requirements
- Mission
  - Proposed mission statement
  - Proposed commander's intent

- Command relationships
- Conclusion—potential resource shortfalls
- Mission analysis approval and commander's course of action planning guidance

Figure V-6. Example Mission Analysis Briefing

(3) Immediately after the mission analysis briefing, the commander approves a restated mission. This can be the staff's recommended mission statement, a modified version of the staff's recommendation, or one that the commander has developed personally. **Once approved, the restated mission becomes the unit mission.**

(4) At the mission analysis brief, the commander will likely describe an updated understanding of the OE, the problem, and the vision of the operational approach to the entire assemblage, which should include representatives from subordinate commands and other partner organizations. This provides the ideal venue for facilitating unity of understanding and vision, which is essential to unity of effort.

---

q. **Publish Commander's Refined Planning Guidance**

(1) After approving the mission statement and issuing the intent, the commander provides the staff (and subordinates in a collaborative environment) with enough additional guidance (including preliminary decisions) to focus the staff and subordinate planning activities during COA development. This refined planning guidance should include the following elements:

(a) An approved mission statement.

(b) Key elements of the OE.

(c) A clear statement of the problem.

(d) Key assumptions.

(e) Key operational limitations.

(f) National strategic objectives with a description of how the operation will support them.

(g) Termination criteria (if appropriate, CCMD-level campaign plans will not have termination criteria and many operations will have transitions rather than termination).

(h) Military objectives or end state and their relation to the national strategic end state.

(i) The JFC's initial thoughts on the conditions necessary to achieve objectives.

(j) Acceptable or unacceptable levels of risk in key areas.

(k) The JFCs visualization of the operational approach to achieve the objectives in broad terms. This operational approach sets the basis for development of COAs. The commander should provide as much detail as appropriate to provide the right level of freedom to the staff in developing COAs. Planning guidance should also address the role of interorganizational and multinational partners in the pending operation and any related special considerations as required.

(2) Commanders describe their visualization of the forthcoming campaign or operations to help build a shared understanding among the staff. Enough guidance (preliminary decisions) must be provided to allow the subordinates to plan the action necessary to accomplish the mission consistent with commander's intent. The commander's guidance must focus on the essential tasks and associated objectives that support the accomplishment of the assigned national objectives. It emphasizes in broad terms when, where, and how the commander intends to employ military capabilities

integrated with other instruments of national power to accomplish the mission within the higher JFC's intent.

(3) The JFC may provide the planning guidance to the entire staff and/or subordinate JFCs or meet each staff officer or subordinate unit individually as the situation and information dictates. The guidance can be given in a written form or orally. No format for the planning guidance is prescribed. However, the guidance should be sufficiently detailed to provide a clear direction and to avoid unnecessary efforts by the staff or subordinate and supporting commands.

(4) Planning guidance can be very explicit and detailed, or it can be very broad, allowing the staff and/or subordinate commands wide latitude in developing subsequent COAs. However, no matter its scope, the content of planning guidance must be arranged in a logical sequence to reduce the chances of misunderstanding and to enhance clarity. Moreover, one must recognize that all the elements of planning guidance are tentative only. The JFC may issue successive planning guidance during the decision-making process; yet the focus of the JFC's staff should remain upon the framework provided in the initial planning guidance. The commander should continue to provide refined planning guidance during the rest of the planning process while understanding of the problem continues to develop.

5. **Course of Action Development (Step 3)**

a. **Introduction**

(1) **A COA is a potential way (solution, method) to accomplish the assigned mission.** The staff develops COAs to provide unique options to the commander, all oriented on accomplishing the military end state. A good COA accomplishes the mission within the commander's guidance, provides flexibility to meet unforeseen events during execution, and positions the joint force for future operations. It also gives components the maximum latitude for initiative.

(2) Figure V-7 shows the key inputs and outputs of COA development. The products of mission analysis drive COA development. Since the operational approach contains the JFC's broad approach to solve the problem at hand, each COA will expand this concept with the additional details that describe **who** will take the action, **what type** of military action will occur, **when** the action will begin, **where** the action will occur, **why** the action is required (purpose), and **how** the action will occur (method of employment of forces). Likewise, **the essential tasks identified during mission analysis (and embedded in the draft mission statement) must be common to all potential COAs.**

(3) Planners can vary COAs by adjusting the use of joint force capabilities throughout the OE by employing the capabilities in combination for effectiveness making use of the information environment (including cyberspace) and the electromagnetic spectrum.

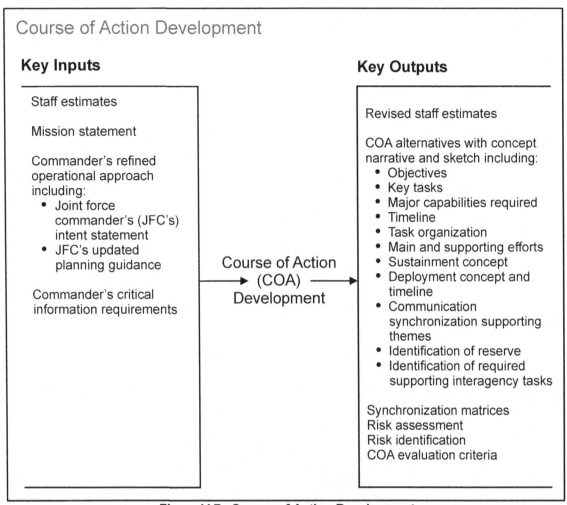

Figure V-7. Course of Action Development

b. **COA Development Considerations**

(1) The products of COA development are potential **COA alternatives,** with a sketch for each if possible. Each COA describes, in broad but clear terms, what is to be done throughout the campaign or operation, the size of forces deemed necessary, time in which joint force capabilities need to be brought to bear, and the risks associated with the COA. These COAs will undergo additional validity testing, analysis, wargaming, and comparison, and they could be eliminated at any point during this process. These COAs provide conceptualization and broad descriptions of potential CONOPS for the conduct of operations that will accomplish the desired end state.

(2) Available planning time is always a key consideration, particularly in a crisis. The JFC gives the staff additional considerations early in COA development to focus the staff's efforts, helping the staff concentrate on developing COAs that are the most appropriate. There should always be more than one way to accomplish the mission, which suggests that commanders and planners should give due consideration to the pros and cons of valid COA alternatives. However, developing several COAs could violate time constraints. Usually, the staff develops two or three COAs to focus their efforts and

concentrate valuable resources on the most likely scenarios. However, COAs must be substantially distinguishable from each other. Commanders should not overburden staffs by developing similar solutions to the problem. The commander's involvement in the early operational design process can help ensure only value-added options are considered. If time and personnel resources permit, different COAs could be developed by different teams to ensure they are unique.

(3) For each COA, the commander must envision the employment of all participants in the operation as a whole—US military forces, MNFs, and interagency and multinational partners—taking into account operational limitations, political considerations, the OA, existing FDOs, and the conclusions previously drawn during the mission analysis, the commander's guidance and informal dialogue and formal IPRs with DOD leadership held to date.

(4) During COA development, the commander and staff consider all feasible adversary COAs. Other actors may also create difficult conditions that must be considered during COA development. It is best to consider all opposing actors' actions likely to challenge the attainment of the desired end states when exploring adversary COAs.

(5) Each COA typically has an associated initial CONOPS with a narrative and sketch and includes the following:

(a) OE.

(b) Objectives.

(c) Key tasks and purpose.

(d) Forces and capabilities required, to include anticipated interagency roles, actions, and supporting tasks.

(e) Integrated timeline.

(f) Task organization.

(g) Operational concept.

(h) Sustainment concept.

(i) Communication synchronization.

(j) Risk.

(k) Required decisions and decision timeline (e.g., mobilization, DEPORD).

(l) Deployment concept.

(m) Main and supporting efforts.

(6) An alternative is an activity within a COA that may be executed to enable achieving an objective. Alternatives, and groups of alternatives comprising branches, allow the commander to act rapidly and transition as conditions change through the campaign or operation. Alternatives, and more broadly branches, should enable the commander to progress sequentially or skip ahead based on success or other changes to the conditions or strategic direction from dialogue with higher commanders, SecDef, and/or the President. They should also enable the commander to transition rapidly, exploit success, and control escalation and tempo while denying the same to the enemy. The development of alternatives within COAs empowers the commander and translates up and down the chain of command and enables strategic flexibility for SecDef and the President. COAs should be simple and brief, yet complete. Individual COAs should have descriptive titles. Distinguishing factors of the COA may suggest titles that are descriptive in nature.

c. **COA Development Techniques and Procedures**

(1) **Review information** contained in the mission analysis and commander's operational approach, planning guidance, and intent statement. All staff members must understand the mission and the tasks that must be accomplished within the commander's intent to achieve mission success.

(2) **Determine the COA Development Technique**

(a) A critical first decision in COA development is whether to conduct simultaneous or sequential development of the COAs. Each approach has distinct advantages and disadvantages. The advantage of simultaneous development of COAs is potential time savings. Separate groups are simultaneously working on different COAs. The disadvantage of this approach is that the synergy of the JPG may be disrupted by breaking up the team. The approach is manpower intensive and requires component and directorate representation in each COA group, and there is an increased likelihood that the COAs will not be distinctive. While there is potential time to be saved, experience has demonstrated that it is not an automatic result. The simultaneous COA development approach can work, but its inherent disadvantages must be addressed and some risk accepted up front. The recommended approach if time and resources allows is the sequential method.

(b) There are several planning sequence techniques available to facilitate COA development. One option is the step-by-step approach (see Figure V-8), which uses the backward-planning technique (also known as reverse planning).

(3) **Review operational objectives and tasks and develop ways to accomplish tasks.** Planners must review and refine theater and supporting operational objectives from the initial work done during the development of the operational approach. These objectives establish the conditions necessary to help accomplish the national strategic objectives. Tasks are shaped by the CONOPS—intended sequencing and integration of air, land, maritime, special operations, cyberspace, and space forces. Tasks are prioritized while considering the enemy's objectives and the need to gain advantage.

Step-by-Step Approach to Course of Action Development

Step   Action

1   Determine how much force will be needed in the theater at the end of the operation or campaign, what those forces will be doing, and how those forces will be postured geographically. Use troop-to-task analysis. Draw a sketch to help visualize the forces and their locations.

2   Looking at the sketch and working backwards, determine the best way to get the forces postured in Step 1 from their ultimate positions at the end of the operation or campaign to a base in friendly territory. This will help formulate the desired basing plan.

3   Using the mission statement as a guide, determine the tasks the force must accomplish en route to their locations/positions at the end of the operation or campaign. Draw a sketch of the maneuver plan. Make sure the force does everything the Secretary of Defense (SecDef) has directed the commander to do (refer to specified tasks from the mission analysis).

4   Determine the basing required to posture the force in friendly territory, and the tasks the force must accomplish to get to those bases. Sketch this as part of the deployment plan.

5   Determine if the planned force is enough to accomplish all the tasks SecDef has given the commander.

6   Given the tasks to be performed, determine in what order the forces should be deployed into theater. Consider the force categories such as combat, protection, sustainment, theater enablers, and theater opening.

7   The information developed should now allow determination of force employment, major tasks and their sequencing, sustainment, and command relationships.

Figure V-8. Step-By-Step Approach to Course of Action Development

(a) Regardless of the eventual COA, the staff should plan to accomplish the higher commander's intent by understanding its essential task(s) and purpose and the intended contribution to the higher commander's mission success.

(b) The staff must ensure all the COAs developed will fulfill the command mission and the purpose of the operation by conducting a review of all essential tasks developed during mission analysis. They should then consider ways to accomplish the other tasks.

(4) Once the staff has begun to visualize COA alternatives, it should see how it can best synchronize (arrange in terms of time, space, and purpose) the actions of all the elements of the force. The staff should estimate the anticipated duration of the operation. One method of synchronizing actions is the use of phasing as discussed earlier. Phasing assists the commander and staff to visualize and think through the entire operation or campaign and to define requirements in terms of forces, resources, time, space, and

purpose. Planners should then **integrate and synchronize** these requirements by using the joint functions of C2, intelligence, fires, movement and maneuver, protection, sustainment, and information. Additionally, planners should consider IRCs as additional tools to create desired effects. At a minimum, planners should make certain the synchronized actions answer the following questions:

(a) How do land, maritime, air, space, cyberspace, and special operations forces integrate across the joint functions to accomplish their assigned tasks?

(b) How can the joint forces synchronize their actions and messages (words and deeds) and integrate IRCs with lethal fires?

(5) The COAs should focus on **COGs and decisive points or areas of influence for CCMD-level campaigns.** The commander and the staff review and refine their COG analysis begun during mission analysis based on updated intelligence, JIPOE products, and initial staff estimates. The refined enemy and friendly COG analysis, particularly the critical vulnerabilities, is considered in the development of the initial COAs. The COG analysis helps the commander become oriented to the enemy and compare friendly strengths and weakness with those of the enemy. By looking at friendly COGs and vulnerabilities, the staff understands the capabilities of their own force and critical vulnerabilities that will require protection. Protection resource limitations will probably mean the staff cannot plan to protect every capability, but rather will look at prioritizing protection for critical capabilities and developing overlapping protection techniques. The strength of one asset or capability may provide protection from the weakness of another.

(6) **Identify the sequencing** (simultaneous, sequential, or a combination) of the actions for each COA. Understand when and what resources become available during the operation or campaign. Resource availability will significantly affect sequencing operations and activities.

For a discussion on defeat and stability mechanisms, see JP 3-0, Joint Operations, *and JP 3-07,* Stability.

(7) **Identify main and supporting efforts** by phase, the purposes of these efforts, and key supporting/supported relationships within phases.

(8) **Identify decision points and assessment process.** The commander will need to know when a critical decision has to be made and how to know specific objectives have been achieved. This requires integration of decision points and assessment criteria into the COA as these processes anticipate a potential need for decisions from outside the command (SecDef, the President, or a functional or adjacent command).

(9) **Identify component-level missions/tasks** (who, what, and where) that will accomplish the stated purposes of main and supporting efforts. Think of component and joint function tasks such as movement and maneuver, intelligence, fires, protection, sustainment, C2, and information. Display them with graphic control measures as much as possible. A designated LOO will help identify these tasks.

(10) **Integrate IRCs.** Some IRCs help to create effects and influence adversary decision making. Planners should consider how IRCs can influence positioning of adversary units, disrupt adversary C2, and decrease adversary morale when developing COAs.

(11) **Task Organization**

(a) The staff should develop an outline task organization to execute the COA. The commander and staff determine appropriate command relationships and appropriate missions and tasks.

(b) **Determine command relationships and organizational options.** Joint force organization and command relationships are based on the operation or campaign CONOPS, complexity, and degree of control required. Establishing **command relationships** includes determining the types of subordinate commands and the degree of authority to be delegated to each. Clear definition of command relationships further clarifies the intent of the commander and contributes to decentralized execution and unity of effort. The commander has the authority to determine the types of subordinate commands from several doctrinal options, including Service components, functional components, and subordinate joint commands. **Regardless of the command relationships selected, it is the JFC's responsibility to ensure these relationships are understood and clear to all subordinate, adjacent, and supporting headquarters.** The following are considerations for establishing joint force organizations:

1. Joint forces will normally be organized with a combination of Service and functional components with operational responsibilities.

2. Functional component staffs should be joint with Service representation in approximate proportion to the mix of subordinate forces. These staffs should be organized and trained prior to employment in order to be efficient and effective, which will require advanced planning.

3. Commanders may establish support relationships between components to facilitate operations.

4. Commanders define the authority and responsibilities of functional component commanders based on the strategic CONOPS and may alter their authority and responsibility during the course of an operation.

5. Commanders must balance the need for centralized direction with decentralized execution.

6. Major changes in the joint force organization are normally conducted at phase changes.

(12) **Sustainment Concept. No COA is complete without a plan to sustain it properly.** The sustainment concept is more than just gathering information on various logistic and personnel services. It entails identifying the requirements for all classes of

supply, creating distribution, transportation, OCS, and disposition plans to support the commander's execution, and organizing capabilities and resources into an overall theater campaign or operation sustainment concept. It concentrates forces and material resources strategically so the right force is available at the designated times and places to conduct decisive operations. It requires thinking through a cohesive sustainment for joint, single Service and supporting forces relationships in conjunction with CSAs, multinational, interagency, nongovernmental, private sector, or international organizations.

(13) **Deployment Concept.** A COA must consider the deployment concept in order to describe the general flow of forces into theater. There is no way to determine the feasibility of the COA without including the deployment concept. While the detailed deployment concept will be developed during plan synchronization, enough of the concept must be described in the COA to visualize force buildup, sustainment requirements, and military-political considerations.

(14) **Define the OA**

(a) The OA is an overarching term that can encompass more descriptive terms for geographic areas. It will provide flexibility/options and/or limitations to the commander. The OA must be precisely defined because the specific geographic area will impact planning factors such as basing, overflight, and sustainment.

(b) OAs include, but are not limited to, such descriptors as AOR, theater of war, theater of operations, JOA, amphibious objective area, joint special operations area, and area of operations. Except for AOR, which is assigned in the UCP, GCCs and their subordinate JFCs designate smaller OAs on a temporary basis. OAs have physical dimensions composed of some combination of air, land, maritime, and space domains. JFCs define these areas with geographical boundaries, which facilitate the coordination, integration, and deconfliction of joint operations among joint force components and supporting commands. The size of these OAs and the types of forces employed within them depend on the scope and nature of the crisis and the projected duration of operations.

See JP 3-0, Joint Operations, *for additional information on OAs.*

(15) **Develop Initial COA Sketches and Statements.** Each COA should answer the following questions:

(a) Who (type of forces) will execute the tasks?

(b) What are the tasks?

(c) Where will the tasks occur? (Start adding graphic control measures, e.g., areas of operation, amphibious objective areas).

(d) When will the tasks begin?

(e) What are key/critical decision points?

(f) How (but do not usurp the components' prerogatives) the commander should provide "operational direction" so the components can accomplish "tactical actions."

(g) Why (for what purpose) will each force conduct its part of the operation?

(h) How will the commander identify successful accomplishment of the mission?

(i) Develop an initial intelligence support concept.

(16) **Test the Validity of Each COA.** All COAs selected for analysis must be valid, and the **staff should reject COA alternatives that do not meet all five of the following validity criteria:**

(a) **Adequate**—Can accomplish the mission within the commander's guidance. Preliminary tests include:

<u>1</u>. Does it accomplish the mission?

<u>2</u>. Does it meet the commander's intent?

<u>3</u>. Does it accomplish all the essential tasks?

<u>4</u>. Does it meet the conditions for the end state?

<u>5</u>. Does it take into consideration the enemy and friendly COGs?

(b) **Feasible**—Can accomplish the mission within the established time, space, and resource limitations.

<u>1</u>. Does the commander have the force structure and lift assets (means) to execute it? The COA is feasible if it can be executed with the forces, support, and technology available within the constraints of the physical domains and against expected enemy opposition.

<u>2</u>. Although this process occurs during COA analysis and the test at this time is preliminary, it may be possible to declare a COA infeasible (for example, resources are obviously insufficient). However, it may be possible to fill shortfalls by requesting support from the commander or other means.

(c) **Acceptable**—Must balance cost and risk with the advantage gained.

<u>1</u>. Does it contain unacceptable risks? (Is it worth the possible cost?) A COA is considered acceptable if the estimated results justify the risks. The basis of this test consists of an estimation of friendly losses in forces, time, position, and opportunity.

<u>2</u>. Does it take into account the limitations placed on the commander (must do, cannot do, other physical or authority limitations)?

<u>3</u>. Acceptability is considered from the perspective of the commander by reviewing the strategic objectives.

<u>4</u>. Are COAs reconciled with external constraints, particularly ROE? This requires visualization of execution of the COA against each enemy capability. Although this process occurs during COA analysis and the test at this time is preliminary, it may be possible to declare a COA unacceptable if it violates the commander's definition of acceptable risk.

(d) **Distinguishable**—Must be sufficiently different from other COAs in the following:

<u>1</u>. The focus or direction of main effort.

<u>2</u>. The scheme of maneuver.

<u>3</u>. Sequential versus simultaneous maneuvers.

<u>4</u>. The primary mechanism for mission accomplishment.

<u>5</u>. Task organization.

<u>6</u>. The use of reserves.

(e) **Complete**—Does it answer the questions who, what, where, when, how, and why? The COA must incorporate:

<u>1</u>. Objectives, desired effects to be created, and tasks to be performed.

<u>2</u>. Major forces required.

<u>3</u>. Concepts for deployment, employment, and sustainment.

<u>4</u>. Time estimates for achieving objectives.

<u>5</u>. Military end state and mission success criteria (including the assessment: how the commander will know they have achieved success).

(17) **Conduct COA Development Brief to Commander.** Figure V-9 provides suggested sequence and content.

(18) **JFC Provides Guidance on COAs**

(a) Review and approve COA(s) for further analysis.

(b) Direct revisions to COA(s), combinations of COAs, or development of additional COA(s).

Example Course of Action Development Briefing

- Operations Directorate of a Joint Staff (J-3)/Plans Directorate of a Joint Staff (J-5)

  ○ Context/background (i.e., road to war)
  ○ Initiation—review guidance for initiation
  ○ Strategic guidance—planning tasks assigned to supported commander, forces/resources apportioned, planning guidance, updates, defense agreements, theater campaign plan(s), Guidance for Employment of the Force/Joint Strategic Campaign Plan
  ○ Forces allocated/assigned

- Intelligence Directorate of a Joint Staff (J-2)

  ○ Joint Intelligence Preparation of the Operational Environment
  ○ Enemy objectives
  ○ Enemy courses of action (COAs)—most dangerous, most likely; strengths and weaknesses

- J-3/J-5

  ○ Update facts and assumptions
  ○ Mission statement
  ○ Commander's intent (purpose, method, end state)
  ○ End state: political/military
    – termination criteria
  ○ Center of gravity analysis results: critical factors; strategic/operational
  ○ Joint operations area/theater of operations/communications zone sketch
  ○ Shaping activities recommended (for current theater campaign plan)
  ○ Flexible deterrent options with desired effect
  ○ For each COA, sketch and statement by phase
    – task organization
    – component tasking
    – timeline
    – recommended command and control by phase
    – lines of operation/lines of effort
    – logistics estimates and feasibility
    – COA risks
    – synchronization matrices
  ○ COA summarized distinctions
  ○ COA priority for analysis

- Update Course of Action Development Briefing to Include:

  ○ Red objectives

- Commander's Guidance

Figure V-9. Example Course of Action Development Briefing

(c) Direct priority for which enemy COA(s) will be used during wargaming of friendly COA(s).

(19) **Continue the Staff Estimate Process.** The staff must continue to conduct their staff estimates of supportability for each COA.

(20) **Conduct Vertical and Horizontal Parallel Planning**

(a) Discuss the planning status of staff counterparts with both commander's and JFC components' staffs.

(b) Coordinate planning with staff counterparts from other functional areas.

(c) Permit adjustments in planning as additional details are learned from higher and adjacent echelons, and permit lower echelons to begin planning efforts and generate questions (e.g., requests for information).

d. **The Planning Directive**

(1) The **planning directive** identifies planning responsibilities for developing joint force plans. It provides guidance and requirements to the staff and subordinate commands concerning coordinated planning actions for plan development. The JFC normally communicates initial planning guidance to the staff, subordinate commanders, and supporting commanders by publishing a planning directive to ensure everyone understands the commander's intent and to achieve unity of effort.

(2) Generally, the plans directorate of a joint staff (J-5) coordinates staff action for planning for the CCMD campaign and contingencies, and the operations directorate of a joint staff (J-3) coordinates staff action in a crisis situation. The J-5 staff receives the JFC's initial guidance and combines it with the information gained from the initial staff estimates. The JFC, through the J-5, may convene a preliminary planning conference for members of the JPEC who will be involved with the plan. This is the opportunity for representatives to meet face-to-face. At the conference, the JFC and selected members of the staff brief the attendees on important aspects of the plan and may solicit their initial reactions. Many potential conflicts can be avoided by this early exchange of information.

6. **Course of Action Analysis and Wargaming (Step 4)**

a. **Introduction**

(1) COA analysis is the process of closely examining potential COAs to reveal details that will allow the commander and staff to tentatively identify COAs that are valid and identify the advantages and disadvantages of each proposed friendly COA. The commander and staff analyze each COA separately according to the commander's guidance. While time-consuming, COA analysis should reaffirm the validity of the COA while answering **'is the COA feasible, and is it acceptable?'**

(2) **Wargaming** is a primary means to conduct this analysis. Wargames are representations of conflict or competition in a synthetic environment, in which people make decisions and respond to the consequences of those decisions. COA wargaming is a conscious attempt to visualize the flow of the operation, given joint force strengths and dispositions, adversary capabilities and possible COAs, the OA, and other aspects of the OE. Each critical event within a proposed COA should be wargamed based upon time available using the action, reaction, and counteraction method of friendly and/or opposing

force interaction. The basic COA wargaming method can be modified to fit the specific mission and OE, and be applied to noncombat, CCMD campaign activities, and combat operations. Wargaming is most effective when it contains the following elements:

(a) People making decisions.

(b) A fair competitive environment (i.e., the game should have no rules or procedures designed to tilt the playing field toward one side or another).

(c) Adjudication.

(d) Consequences of actions.

(e) Iterative (i.e., new insights will be gained as games are iterated).

(3) COA wargaming allows the commander, staff, and subordinate commanders and their staffs to gain a common understanding of friendly and enemy COAs, and other actor actions that may (intentionally or otherwise) work in opposition to achieving the objectives or attaining desired end state conditions. This common understanding allows them to determine the advantages and disadvantages of each COA and forms the basis for the commander's comparison and approval. COA wargaming involves a detailed evaluation of each COA as it pertains to the enemy and the OE. Each of the selected friendly COAs is then wargamed against selected enemy or OE COAs, as well as other actor actions as applicable (for example, wargaming theater campaign or functional campaign activities can identify how a HN or third party might react/respond to US campaign activities). The commander will select the COAs he wants wargamed and provide wargaming guidance along with refined evaluation criteria.

(4) Wargaming stimulates thought about the operation so the staff can obtain ideas and insights that otherwise might not have emerged. An objective, comprehensive analysis of COA alternatives is difficult even without time constraints. Based upon time available, the commander should wargame each COA alternative against the most probable and the most dangerous adversary COAs (or most difficult objectives in noncombat and campaign operations) identified through the JIPOE process. Figure V-10 shows the key inputs and outputs associated with COA analysis.

b. **Analysis and Wargaming Process**

(1) The analysis and wargaming process can be as simple as a detailed narrative effort that describes the action, probable reaction, counteraction, assets, and time used. A more comprehensive version is the "sketch-note" technique, which adds operational sketches and notes to the narrative process in order to gain a clearer picture. Sophisticated wargames employ more extensive means to depict the range of actions by competitors and the consequences of the synthesis of those actions. The most sophisticated form of wargaming is one where all competitors in a conflict are represented (and emulated to the best degree possible) and have equal decision space to enable a full exploration of the competition within the OE. Modeling and simulation are distinct and separate analytic tools and not the same as wargames. Modeling and simulation can be complementary and

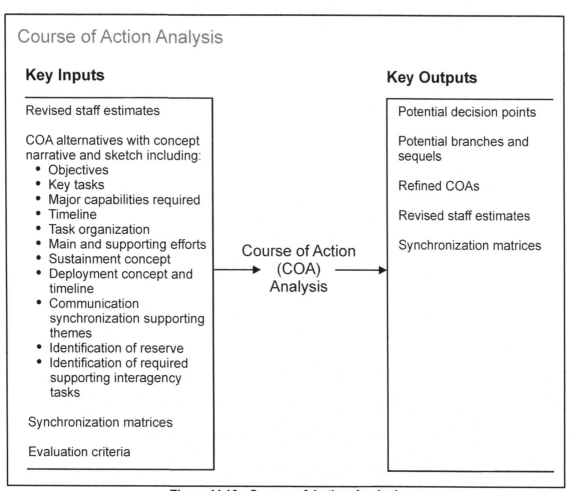

Figure V-10. Course of Action Analysis

assist wargaming through bookkeeping, visualization, and adjudication for well understood actions. The heart of the commander's estimate process is analysis of multiple COAs. The items selected for wargaming and COA comparison will depend on the nature of the mission. For plans or orders involving combat operations, the staff considers opposing COAs based on enemy capabilities, objectives, an estimate of the enemy's intent, and integrated actions by other actors (neutral, other adversaries, and even friendly actions that would not be favorable) that would challenge achievement of the objective. For noncombat operations or CCMD campaign plans, the staff may analyze COAs based on partner capabilities, partner and US objectives, criticality, and risk. In the analysis and wargaming step, the staff analyzes the probable effect each opposing COA has on the chances of success of each friendly COA. The aim is to develop a sound basis for determining the feasibility and acceptability of the COAs. Analysis also provides the planning staff with a greatly improved understanding of their COAs and the relationship between them. COA analysis identifies which COA best accomplishes the mission while best positioning the force for future operations. It also helps the commander and staff to:

(a) Determine how to maximize combat power against the enemy while protecting the friendly forces and minimizing collateral damage in combat or maximize the effect of available resources toward achieving CCMD and national objectives in noncombat operations and campaigns.

(b) Have as near an identical visualization of the operation as possible.

(c) Anticipate events in the OE and potential reaction options.

(d) Determine conditions and resources required for success while also identifying gaps and seams.

(e) Determine when and where to apply the force's capabilities.

(f) Plan for and coordinate authorities to integrate IRCs early.

(g) Focus intelligence collection requirements.

(h) Determine the most flexible COA.

(i) Identify potential decision points.

(j) Determine task organization options.

(k) Develop data for use in a synchronization matrix or related tool.

(l) Identify potential plan branches and sequels.

(m) Identify high-value targets.

(n) Assess risk.

(o) Determine COA advantages and disadvantages.

(p) Recommend CCIRs.

(q) Validate end states and objectives.

(r) Identify contradictions between friendly COAs and expected enemy end states.

(2) Wargaming is a disciplined process, with rules and steps that attempt to visualize the flow of the operation. The process considers friendly dispositions, strengths, and weaknesses; enemy assets and probable COAs; and characteristics of the physical environment. It relies heavily on joint doctrinal foundation, tactical judgment, and operational and regional/area experience. It focuses the staff's attention on each phase of the operation in a logical sequence. It is an iterative process of action, reaction, and counteraction. Wargaming stimulates ideas and provides insights that might not otherwise be discovered. It highlights critical tasks and provides familiarity with operational possibilities otherwise difficult to achieve. Wargaming is a critical portion of the planning process and should be allocated significant time. **Each retained COA should, at a**

minimum, be wargamed against both the most likely and most dangerous enemy COAs.

(3) During the wargame, the staff takes a COA statement and begins to add more detail to the concept, while determining the strengths or weaknesses of each COA. Wargaming tests a COA and can provide insights that can be used to improve upon a developed COA. The commander and staff (and subordinate commanders and staffs if the wargame is conducted collaboratively) may change an existing COA or develop a new COA after identifying unforeseen critical events, tasks, requirements, or problems.

(4) For the wargame to be effective, the commander should indicate what aspects of the COA should be examined and tested. Wargaming guidance should include the list of friendly COAs to be wargamed against specific threat COAs (e.g., COAs 1, 2, and 3 against the enemy's most likely and most dangerous COAs), the timeline for the phase or stage of the operations, a list of critical events, and level of detail (i.e., two levels down). In order for a valid COA comparison (JPP step 5), each friendly COA must be wargamed against the same set of threat COAs.

(5) **COA Analysis Considerations.** Evaluation criteria and known critical events are two of the many important considerations as COA analysis begins.

(a) The commander and staff use evaluation criteria during follow-on COA comparison (JPP step 5) for the purpose of selecting the best COA. The commander and staff consider various potential evaluation criteria during wargaming and select those that the staff will use during COA comparison to evaluate the effectiveness and efficiency of one COA relative to others following the wargame. These evaluation criteria help focus the wargaming effort and provide the framework for data collection by the staff. These criteria are those aspects of the situation (or externally imposed factors) that the commander deems critical to mission accomplishment. Figure V-11 shows examples of potential evaluation criteria.

For more information, see Appendix G, "Course of Action Comparison."

(b) Evaluation criteria change from mission to mission. It will be helpful during future wargaming steps for all participants to be familiar with the criteria so any insights that influence a criterion are recorded for later comparison. The criteria may include anything the commander desires. If they are not received directly, the staff can derive them from the commander's intent statement. Evaluation criteria do not stand alone. Each must be clearly defined. Precisely defining criteria reduces subjectivity and ensures consistent evaluation. The following sources provide a good starting point for developing a list of potential evaluation criteria.

<u>1.</u> Commander's guidance and commander's intent.

<u>2.</u> Mission accomplishment at an acceptable cost.

<u>3.</u> The principles of joint operations.

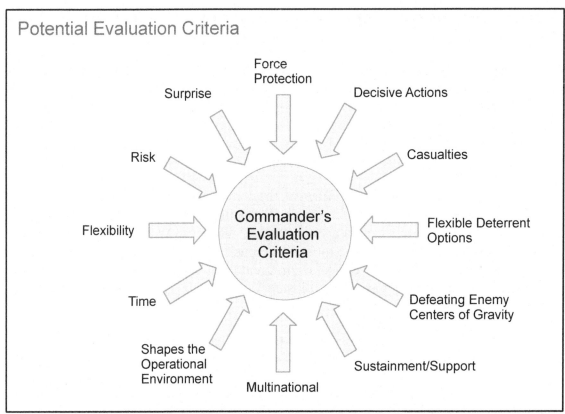

Figure V-11. Potential Evaluation Criteria

4. Doctrinal fundamentals for the type of operation being conducted.

5. The level of residual risk in the COA.

6. Implicit significant factors relating to the operation (e.g., need for speed, security).

7. Factors relating to specific staff functions.

8. Elements of operational design.

9. Other factors to consider: diplomatic or political constraints, residual risks, financial costs, flexibility, simplicity, surprise, speed, mass, sustainability, C2, and infrastructure survivability.

(c) **List Known Critical Events.** These are essential tasks, or a series of critical tasks, conducted over a period of time that require detailed analysis (such as the series of component tasks to be performed on D-day).

1. This may be expanded to review component tasks over a phase(s) of an operation or over a period of time (C-day through D-day). The planning staff may wish at this point to also identify decision points (those decisions in time and space that the commander must make to ensure timely execution and synchronization of resources).

These decision points are most likely linked to a critical event (e.g., commitment of the reserve force).

2. For CCMD campaigns, this includes identifying linked events and activities: the staff must identify if campaign activities are sensitive to the sequence in which they are executed and if subsequent activities are dependent on the success of earlier ones. If resources are cut for an activity early in the campaign, the staff must identify to the commander the impact of the loss of that event (or if the results were different from those anticipated), a decision point to continue subsequent events, and alternates if the planned events were dependent on earlier ones.

(6) **There are two key decisions to make before COA analysis begins.** The first decision is to decide what type of wargame will be used. This decision should be based on commander's guidance, time and resources available, staff expertise, and availability of simulation models. The second decision is to prioritize the enemy COAs or the partner capabilities, partner and US objectives for noncombat operations, and the wargame that it is to be analyzed against. In time-constrained situations, it may not be possible to wargame against all COAs.

c. **Conducting the Wargame**

(1) The primary steps are: prepare for the wargame, conduct the wargame, evaluate the results, and prepare products. Figure V-12 shows sample wargaming steps.

(2) **Prepare for the Wargame**

(a) The two forms of wargames are computer-assisted and manual. There are many forms of computer-assisted wargames; most require a significant amount of preparation to develop and load scenarios and then to train users. However, the potential to utilize the computer model for multiple scenarios or blended scenarios makes it valuable. For both types, consider how to organize the participants in a logical manner.

(b) For manual wargaming, three distinct methods are available to run the event:

1. **Deliberate Timeline Analysis.** Consider actions day-by-day or in other discrete blocks of time. This is the most thorough method for detailed analysis when time permits.

2. **Phasing.** Used as a framework for COA analysis. Identify significant actions and requirements by functional area and/or joint task force (JTF) component.

3. **Critical Events/Sequence of Essential Tasks.** The sequence of essential tasks, also known as the critical events method, highlights the initial actions necessary to establish the conditions for future actions, such as a sustainment capability and engage enemy units in the deep battle area. At the same time, it enables the planners to adapt if the enemy or other actor in the OE reacts in such a way that necessitates

Sample Wargaming Steps

1 **Prepare for the Wargame**
- Gather tools
- List and review friendly forces and capabilities
- List and review opposing forces and capabilities
- List known critical events
- Determine participants
- Determine opposing course of action (COA) to wargame
- Select wargaming method
  - manual or computer-assisted
- Select a method to record and display wargaming results
  - narrative
  - sketch and note
  - war game worksheets
  - synchronization matrix

2 **Conduct Wargame and Assess Results**
- Purpose of wargame (identify gaps, visualization, etc.)
- Basic methodology (e.g., action, reaction, counteraction)
- Record results

3 **Prepare Products**
- Results of the wargame brief
  - potential decision points
  - evaluation criteria
  - potential branches and sequels
- Revised staff estimates
- Refined COAs
- Time-phased force and deployment data refinement and transportation feasibility
- Feedback through the COA decision brief

Figure V-12. Sample Wargaming Steps

reordering of the essential tasks. This technique also allows wargamers to analyze concurrently the essential tasks required to execute the CONOPS. Focus on specific critical events that encompass the essence of the COA. If necessary, different MOEs should be developed for assessing different types of critical events (e.g., destruction, blockade, air control, neutralization, ensure defense). As with the focus on phasing, the critical events discussion identifies significant actions and requirements by functional area and/or by JTF component and enables a discussion of possible or expected reactions to execution of critical tasks.

(c) **Red Cell.** The J-2 staff will provide a red cell to role-play and model the enemies and other relevant actors in the OE during planning and specifically during wargaming.

<u>1</u>. A robust, well-trained, imaginative, and skilled red cell that aggressively pursues the enemy's point of view during wargaming is essential. By accurately portraying the full range of realistic capabilities and options available to the enemy, they help the staff address friendly responses for each enemy COA. For campaign

and noncombat operation planning, the red cell provides expected responses to US actions based on their knowledge and analysis of the OE.

2. The red cell is normally composed of personnel from the joint force J-2 staff and augmented by other subject matter experts.

3. The red cell develops critical decision points, projects enemy and other actor's OE reactions to friendly actions, and estimates impacts and implications on the enemy forces and objectives. By trying to win the wargame, the red cell helps the staff identify weaknesses and vulnerabilities before a real enemy does.

4. Given time constraints, as a minimum, the most dangerous and most likely COAs should be wargamed and role-played by the red cell during the wargame.

(d) **White Cell.** A small cell of arbitrators normally composed of senior individuals familiar with the plan is a smart investment to ensure the wargame does not get bogged down in unnecessary disagreement or arguing. The white cell will provide overall oversight to the wargame and any adjudication required between participants. The white cell may also include the facilitator and/or highly qualified experts as required.

(e) In addition to a red cell and a white cell, there should also be a blue cell that represents friendly forces, and a green cell represents transnational groups, NGOs, and neutral regional populations.

(3) **Conduct the Wargame and Evaluate the Results**

(a) The facilitator and the red cell chief get together to agree on the rules of the wargame. The wargame begins with an event designated by the facilitator. It could be an enemy offensive or defensive action, a friendly offensive or defensive action, or some other activity such as a request for support or campaign activity. They decide where (in the OA) and when (H-hour or L-hour) it will begin. They review the initial array of forces and the OE. Of note, they must come to an agreement on the effectiveness of capabilities and previous actions by both sides prior to the wargame. The facilitator must ensure all members of the wargame know what events will be wargamed and what techniques will be used. This coordination within the friendly team and between the friendly and the red team should be done well in advance.

(b) Each COA wargame has a number of turns, each consisting of three total moves: action, reaction, and counteraction. If necessary, each turn of the wargame may be extended beyond the three basic moves. The facilitator, based on JFC guidance, decides how many total turns are made in the wargame.

(c) During the wargame, the participants must continually evaluate the COA's feasibility. Can it be supported? Can this be done? Will it achieve the desired results? Are more forces, resources, intelligence collection capabilities, or time needed? Are necessary logistics and communications available? Is the OA large enough? Has the threat successfully impacted key enablers like logistics or communications, or countered a certain phase or stage of a friendly COA? Based on the answers to the above questions, revisions to the friendly COA may be required. Major revisions to a COA are not made in the midst of a wargame. Instead, stop the wargame, make the revisions, and start over at the beginning.

(d) The wargame is for comparing and contrasting friendly COAs with the enemy COAs. Planners compare and contrast friendly COAs with each other in the fifth step of JPP, COA comparison. Planners avoid becoming emotionally attached to a friendly COA and avoid comparing one friendly COA with another friendly COA during the wargame so they can remain unbiased. The facilitator ensures adherence to the timeline. A wargame for one COA at the JTF level may take six to eight hours. The facilitator must allocate enough time to ensure the wargame will thoroughly test a COA.

(e) A **synchronization matrix** is a decision-making tool and a method of recording the results of wargaming. Key results that should be recorded include decision points, potential evaluation criteria, CCIRs, COA adjustments, branches, and sequels. Using a synchronization matrix helps the staff visually synchronize the COA across time and space in relation to the enemy's possible COAs and (or) other actor's activities within the OA. The wargame and synchronization matrix efforts will be particularly useful in identifying cross-component support resource requirements.

(f) The wargame considers friendly dispositions, strengths, and weaknesses; enemy assets and probable COAs; and characteristics of the OA. Through a logical sequence, it focuses the participants on essential tasks to be accomplished.

(g) When the wargame is complete and the worksheet and synchronization matrix are filled out, there should be enough detail to flesh out the bones of the COA and begin orders development (once the COA has been selected by the commander in a later JPP step).

(h) Additionally, the wargame will produce a refined event template and the initial decision support template (DST), decision points (and the CCIR related to them), or other decision support tools. These are similar to a football coach's game plan. The tools can help predict what the threat will do and how partner nations or other actors will react to US actions. The tools also provide the commander options for employing forces to counter an adversary action. The tools will prepare the commander (coach) and the staff (team) for a wide range of possibilities and a choice of immediate solutions.

(i) The wargame relies heavily on doctrinal foundation, tactical and operational judgment, and experience. It generates new ideas and provides insights that might have been overlooked. The dynamics of the wargame require the red cell to be aggressive, but realistic, in the execution of threat activities. The wargame:

<u>1.</u> Records advantages and disadvantages of each COA as they become evident.

<u>2.</u> Creates decision support tools (a game plan).

<u>3</u>. Focuses the planning team on the threat and commander's evaluation criteria.

(4) **Prepare Products.** Certain products should result from the wargame in addition to wargamed COAs. Planners enter the wargame with a rough event template and must complete the wargame with a refined, more accurate event template. The event template with its named areas of interest (NAIs) and time-phase lines will help the J-2 focus the intelligence collection effort. An event matrix can be used as a "script" for intelligence reporting during the wargame. It can also tell planners if they are relying too much on one or two collection platforms and if assets have been overextended.

(a) A first draft of a DST should also come out of the COA wargame. As more information about friendly forces and threat forces becomes available, the DST may change.

(b) The critical events are associated with the essential tasks identified in mission analysis. The decision points are tied to points in time and space when and where the commander must make a critical decision. Decision points should be tied to the CCIRs. CCIRs generate two types of information requirements: PIRs and FFIRs. The commander approves CCIRs. From a threat perspective, PIRs tied to a decision point will require an intelligence collection plan that prioritizes and tasks collection assets to gather information about the threat. JIPOE ties PIRs to NAIs, which are linked to adversary COAs. The synchronization matrix is a tool that will help determine if adequate resources are available. Primary outputs are:

<u>1</u>. Wargamed COAs with graphic and narrative. Branches and sequels identified.

<u>2</u>. Information on commander's evaluation criteria.

<u>3</u>. Initial task organization.

<u>4</u>. Critical events and decision points.

<u>5</u>. Newly identified resource shortfalls.

<u>6</u>. Refined/new CCIRs and event template/matrix.

<u>7</u>. Initial DST/DSM.

<u>8</u>. Refined synchronization matrix.

<u>9</u>. Refined staff estimates.

<u>10</u>. Assessment plan and criteria.

(c) The outputs of the COA wargame will be used in the JPP steps COA comparison, COA approval, and plan or order development. The results of the wargame are an understanding of the strengths and weaknesses of each friendly COA, the core of the back brief to the commander.

(d) The commander and staff normally will compare advantages and disadvantages of each COA during COA comparison. However, if the suitability, feasibility, or acceptability of any COA becomes questionable during the analysis step, the commander should modify or discard it and concentrate on other COAs. The need to create additional combinations of COAs may also be identified.

7. Course of Action Comparison (Step 5)

a. Introduction

(1) COA comparison is a subjective process whereby COAs are considered independently and evaluated/compared against a set of criteria that are established by the staff and commander. The objective is to identify and recommend the COA that has the highest probability of accomplishing the mission.

(2) Figure V-13 depicts inputs and outputs for COA comparison. Other products not graphically shown in the chart include updated JIPOE products, updated CCIRs, staff estimates, and commander's ID of branches for further planning.

(3) COA comparison facilitates the commander's decision-making process by balancing the **ends, means, ways, and risk** of each COA. The end product of this task is a briefing to the commander on a COA recommendation and a decision by the commander. COA comparison helps the commander answer the following questions:

(a) What are the differences between each COA?

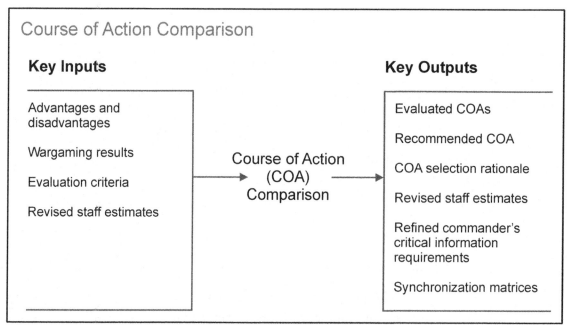

Figure V-13. Course of Action Comparison

(b) What are the advantages and disadvantages?

(c) What are the risks?

b. **COA Comparison Process**

(1) In COA comparison, the staff determines which COA performs best against the established evaluation criteria. The commander reviews the criteria list and adds or deletes, as required. The number of evaluation criteria will vary, but there should be enough to differentiate COAs. COAs are not compared with each other within any one criterion, but rather they are individually evaluated against the criteria that are established by the staff and commander. Their individual performances are then compared to enable the staff to recommend a preferred COA to the commander.

(2) Staff officers may each use their own matrix, such as the example in Figure V-14, to compare COAs with respect to their functional areas. Matrices use the evaluation criteria developed before the wargame. Decision matrices alone cannot provide decision solutions. Their greatest value is providing a method to compare COAs against criteria that the commander and staff believe will produce mission success. They are analytical tools that staff officers use to prepare recommendations. Commanders provide the solution by applying their judgment to staff recommendations and making a decision.

(3) The staff helps the commander identify and select the COA that best accomplishes the mission. The staff supports the commander's decision-making process by clearly portraying the commander's options and recording the results of the process.

Example of Staff Estimate Matrix (Intelligence Estimate)

Evaluation Criteria	Frontal Course of Action 1	Envelopment Course of Action 2
Effects of Terrain		✓
Effects of Weather	✓	
Utilizes Surprise		✓
Attacks Critical Vulnerabilities		✓
Collection Support		✓
Counterintelligence	✓	
Totals	2	4

Figure V-14. Example of Staff Estimate Matrix (Intelligence Estimate)

The staff evaluates feasible COAs to identify the one that performs best within the evaluation criteria against the enemy's most likely and most dangerous COAs.

(4) **Prepare for COA Comparison.** The commander and staff use the evaluation criteria developed during mission analysis to identify the advantages and disadvantages of each COA. Comparing the strengths and weaknesses of the COAs identifies their advantages and disadvantages relative to each other.

(a) Determine/define comparison/evaluation criteria. As discussed earlier, criteria are based on the particular circumstances and should be relative to the situation. There is no standard list of criteria, although the commander may prescribe several core criteria that all staff directors will use. Individual staff sections, based on their estimate process, select the remainder of the criteria.

1. Criteria are based on the particular circumstances and should be relative to the situation.

2. Review commander's guidance for relevant criteria.

3. Identify implicit significant factors relating to the operation.

4. Each staff identifies criteria relating to that staff function.

5. Other criteria might include:

a. Political, social, and safety constraints; requirements for coordination with embassy/interagency personnel.

b. Fundamentals of joint warfare.

c. Elements of operational design.

d. Mission accomplishment.

e. Risks.

f. Costs.

g. Time.

(b) Define and determine the standard for each criterion.

1. Establish standard definitions for each evaluation criterion. Define the criteria in precise terms to reduce subjectivity and ensure the interpretation of each evaluation criterion remains constant between the various COAs.

2. Establish definitions prior to commencing COA comparison to avoid compromising the outcome.

3. Apply standards for each criterion to each COA.

(c) The staff evaluates COAs using those evaluation criteria most important to the commander to identify the one COA with the highest probability of success. The selected COA should also:

1. Place the force in the best posture for future operations.

2. Provide maximum latitude for initiative by subordinates.

3. Provide the most flexibility to meet unexpected threats and opportunities.

c. **Determine the comparison method and record.** Actual comparison of COAs is critical. The staff may use any technique that facilitates reaching the best recommendation and the commander making the best decision. There are a number of techniques for comparing COAs. Examples of several decision matrices can be found in Appendix G, "Course of Action Comparison."

d. **COA comparison is subjective and should not be turned into a strictly mathematical process.** The key is to inform the commander why one COA is preferred over the others in terms of the evaluation criteria and the risk.

8. Course of Action Approval (Step 6)

a. **Introduction**

(1) In this JPP step, the staff briefs the commander on the COA comparison and the analysis and wargaming results, including a review of important supporting information. The staff determines the preferred COA to recommend to the commander. Figure V-15 depicts the COA approval inputs and outputs.

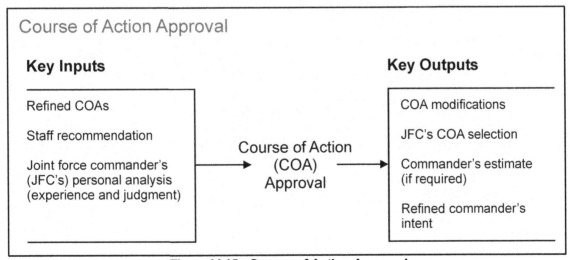

Figure V-15. Course of Action Approval

(2) The nature of the OE or contingency may make it difficult to determine the desired end state until the crisis actually occurs. In these cases, the JFC may choose to present two or more valid COAs for approval by higher authority. A single COA can then be approved when specific circumstances become clear. However, in a crisis, the desired end state should be based on the set of objectives approved by the President.

b. **Prepare and Present the COA Decision Briefing.** The staff briefs the commander on the COA comparison, COA analysis, and wargaming results. The briefing should include a review of important supporting information such as the current status of the joint force, the current JIPOE, and assumptions used in COA development. All principal staff directors and the component commanders should attend this briefing (physically or virtually). Figure V-16 shows a sample COA briefing guide.

c. **Commander Selects/Modifies the COA.** COA selection is the end result of the COA comparison process. Throughout the COA development process, the commander conducts an independent analysis of the mission, possible COAs, and relative merits and risks associated with each COA. The commander, upon receiving the staff's recommendation, combines personal analysis with the staff recommendation, resulting in a selected COA. It gives the staff a concise statement of how the commander intends to accomplish the mission, and provides the necessary focus for planning and plan development. During this step, the commander should:

(1) Review staff recommendations.

(2) Apply results of own COA analysis and comparison.

(3) Consider any separate recommendations from supporting and subordinate commanders.

(4) Review guidance from the higher headquarters/strategic guidance.

(5) The commander may:

(a) Concur with staff/component recommendations, as presented.

(b) Concur with recommended COAs, but with modifications.

(c) Select a different COA from the staff/component recommendation.

(d) Combine COAs to create a new COA.

(e) Reject all and start over with COA development or mission analysis.

(f) Defer the decision and consult with selected staff/commanders prior to making a final decision.

Sample Course of Action Briefing Guide

- Purpose of the briefing
- Opposing situation
  - **Strength.** A review of opposing forces, both committed and available for reinforcement
  - **Composition.** Order of battle, major weapons systems, and operational characteristics
  - **Location and disposition.** Ground combat and fire support forces; air, naval, and missile forces; logistics forces and nodes; command and control facilities; and other combat power
  - **Reinforcements.** Land; air; naval; missile; chemical, biological, radiological, and nuclear; other advanced weapons systems; capacity for movement of these forces
  - **Logistics.** Summary of opposing forces ability to support combat operations
  - **Time and space factors.** The capacity to move and reinforce positions
  - **Combat efficiency.** The state of training, readiness, battle experience, physical condition, morale, leadership, motivation, tactical doctrine, discipline, and significant strengths and weaknesses

- Friendly situation (similar elements as opposing situation)
- Mission statements
- Commander's intent statement
- Operational concepts and courses of action (COAs)
  - Any changes from the mission analysis briefing in the following areas:
    - assumptions
    - limitations
    - adversary and friendly centers of gravity
    - phasing of the operation (if phased)
    - lines of operation/lines of effort
  - Present COAs. As a minimum, discuss:
    - COA# _____ (short name, e.g., "Simultaneous Assault")
    - COA statement (brief concept of operations)
    - COA sketch
    - COA architecture
      - task organization
      - command relationships
      - organization of the operational area
    - major differences between each COA
    - summaries of COAs
  - COA analysis
    - review of the joint planning group's wargaming efforts
    - add considerations from own experiences
  - COA comparisons
    - description of comparison criteria (e.g., evaluation criteria) and comparison methodology
    - weigh strengths and weaknesses with respect to comparison criteria
  - COA recommendations
    - staff
    - components

Figure V-16. Sample Course of Action Briefing Guide

d. **Refine Selected COA.** Once the commander selects a COA, the staff will begin the refinement process of that COA into a clear decision statement to be used in the commander's estimate. At the same time, the staff will apply a final "acceptability" check.

---

(1) Staff refines commander's COA selection into clear decision statement.

(a) Develop a brief statement that **clearly and concisely** sets forth the COA selected and provides whatever information is necessary to develop a plan for the operation (no defined format).

(b) Describe what the force is to do as a whole, and as much of the elements of when, where, and how as may be appropriate.

(c) Express decision in terms of what is to be accomplished, if possible.

(d) Use simple language so the meaning is unmistakable.

(e) Include statement of what is acceptable risk.

(2) Apply final "acceptability" check.

(a) Apply experience and an understanding of situation.

(b) Consider factors of acceptable risk versus desired objectives consistent with higher commander's intent and concept. Determine if gains are worth expenditures.

e. **Prepare the Commander's Estimate**

(1) Once the commander selects the COA, provides guidance, and updates intent, the staff then completes the commander's estimate. The commander's estimate provides a **concise narrative statement** of how the commander intends to accomplish the mission and provides the necessary focus for campaign planning and contingency plan development. Further, it responds to the establishing authority's requirement to develop a plan for execution. The commander's estimate provides a continuously updated source of information from the perspective of the commander. Commanders at various levels use estimates during JPP to support COA determination and plan or order development.

(2) A commander uses a commander's estimate as the situation dictates. The commander's initial intent statement and planning guidance to the staff can provide sufficient information to guide the planning process. The commander will tailor the content of the commander's estimate based on the situation and ongoing analysis. A typical format for a commander's estimate is in CJCSM 3130.03, *Adaptive Planning and Execution (APEX) Planning Formats and Guidance.*

(a) Contents may vary, depending on the nature of the plan or contingency, time available, and the applicability of prior planning. In a rapidly developing situation, the formal commander's estimate may be impractical, and the entire estimate process may be reduced to a commanders' conference.

(b) With appropriate horizontal and vertical coordination, the commander's COA selection may be briefed to and approved by SecDef. In the strategic context, where military operations are strategically significant, even a commander's selected COA is normally briefed to and approved by the President or SecDef. The commander's estimate then becomes a matter of formal record keeping and guidance for component and supporting forces.

(3) The supported commander may use simulation and analysis tools in the collaborative environment to evaluate a variety of options, and may also choose to convene a concept development conference involving representatives of subordinate and supporting commands, the Services, JS, and other interested parties. Review of the resulting commander's estimate requires collaboration and coordination among all planning participants. The supported commander may highlight issues for future interagency consultation, review, or resolution to be presented to SecDef during the IPR.

(4) **CJCS Estimate Review.** The estimate review determines whether the scope and concept of planned operations satisfy the tasking and will accomplish the mission, determines whether the assigned tasks can be accomplished using available resources in the timeframes contemplated by the plan, and ensures the plan is proportional and worth the expected costs. As planning is approved by SecDef (or designated representative) during an IPR, the commander's estimate informs the refinement of the initial CONOPS for the plan.

9. Plan or Order Development (Step 7)

a. CONOPS

(1) The CONOPS clearly and concisely expresses what the JFC intends to accomplish and how it will be done using available resources. It describes how the actions of the joint force components and supporting organizations will be integrated, synchronized, and phased to accomplish the mission, including potential branches and sequels. The CONOPS:

(a) States the commander's intent.

(b) Describes the central approach the JFC intends to take to accomplish the mission.

(c) Provides for the application, sequencing, synchronization, and integration of forces and capabilities in time, space, and purpose (including those of multinational and interagency organizations as appropriate).

(d) Describes when, where, and under what conditions the supported commander intends to conduct operations and give or refuse battle, if required.

(e) Focuses on friendly, allied, partner, and adversary COGs and their associated critical vulnerabilities.

(f) Provides for controlling the tempo of the operation.

(g) Visualizes the campaign in terms of the forces and functions involved.

(h) Relates the joint force's objectives and desired effects to those of the next higher command and other organizations as necessary. This enables assignment of tasks to subordinate and supporting commanders.

(2) **Planning results in a plan that is documented in the format of a plan or an order. If execution is imminent or in progress, the plan is typically documented in the format of an order.** During plan or order development, the commander and staff, in collaboration with subordinate and supporting components and organizations, expand the approved COA into a detailed plan or OPORD by refining the initial CONOPS associated with the approved COA. **The CONOPS is the centerpiece of the plan or OPORD.**

(3) The staff writes (or graphically portrays) the CONOPS in sufficient detail so subordinate and supporting commanders understand their mission, tasks, and other requirements and can develop their supporting plans. During CONOPS development, the commander determines the best arrangement of simultaneous and sequential actions and activities to accomplish the assigned mission consistent with the approved COA, and resources and authorities available. This arrangement of actions dictates the sequencing of activities or forces into the OA, providing the link between the CONOPS and force planning. The link between the CONOPS and force planning is preserved and perpetuated through the sequencing of forces into the OA via a TPFDD. The structure must ensure unit integrity, force mobility, and force visibility as well as the ability to transition to branches or sequels rapidly as operational conditions dictate. Planners ensure the CONOPS, force plan, deployment plans, and supporting plans provide the flexibility to adapt to changing conditions, and are consistent with the JFC's intent.

(4) If the scope, complexity, and duration of the military action contemplated to accomplish the assigned mission warrants execution via a series of related operations, then the staff outlines the CONOPS as a campaign. They develop the preliminary part of the operational campaign in sufficient detail to impart a clear understanding of the commander's concept of how the assigned mission will be accomplished.

(5) During CONOPS development, the JFC must assimilate many variables under conditions of uncertainty to determine the essential military conditions, sequence of actions, and application of capabilities and associated forces to create effects and achieve objectives. **JFCs and their staffs must be continually aware of the higher-level objectives and associated desired and undesired effects that influence planning at every juncture.** If operational objectives are not linked to strategic objectives, the inherent linkage or "nesting" is broken and eventually tactical considerations can begin to drive the overall strategy at cross-purposes.

CJCSM 3130.03, Adaptive Planning and Execution (APEX) Planning Formats and Guidance, *provides detailed guidance on CONOPS content and format.*

b. **Format of Military Plans and Orders.** Plans and orders can come in many varieties from very detailed campaign plans and contingency plans to simple verbal orders. They may also include orders and directives such as OPORDs, WARNORDs, PLANORDs, ALERTORDs, EXORDs, and FRAGORDs, as well as PTDOs, DEPORDs, and the GFMAP. The more complex directives will contain much of the amplifying information in appropriate annexes and appendices. However, the directive should always contain the essential information in the main body. The information contained may depend on the time available, the complexity of the operation, and the levels of command involved. In most cases, the directive will be standardized in the five-paragraph format that is described in CJCSM 3130.03, *Adaptive Planning and Execution (APEX) Planning Formats and Guidance.* Following is a brief description of each of these paragraphs.

(1) **Paragraph 1—Situation.** The commander's summary of the general situation that ensures subordinates understand the background of the planned operations. Paragraph 1 will often contain subparagraphs describing the higher commander's intent, friendly forces, and enemy forces.

(2) **Paragraph 2—Mission.** The commander's mission statement.

(3) **Paragraph 3—Execution.** This paragraph contains commander's intent, which will enable commanders two levels down to exercise initiative while keeping their actions aligned with the overall purpose of the mission. It also specifies objectives, tasks, and assignments for subordinates (by phase, as applicable—with clear criteria denoting phase transition and completion).

(4) **Paragraph 4—Administration and Logistics.** This paragraph describes the concept of support for logistics, personnel, and health services.

(5) **Paragraph 5—C2.** This paragraph specifies the command relationships, succession of command, and overall plan for communications.

c. **Plan or Order Development**

(1) For most plans and orders, the CJCS monitors planning activities, resolves shortfalls when required, and reviews the supported commander's plan for adequacy, feasibility, acceptability, completeness, and compliance with policy and joint doctrine. When required, the commander will conduct one or more IPRs with SecDef (or designated representative) to confirm the plan's strategic guidance, assumptions (including timing and national-level decisions required), any limitations (restrictions and constraints), the mission statement, the operational approach, key capability shortfalls, areas of risk, and acceptable levels of risk; and any further guidance required for plan refinement. During the IPRs, the CJCS and the USD(P) will separately address issues arising from, or resolved during, plan review (e.g., key risks, decision points). Commanders should show how the plan supports the objectives identified in the GEF and JSCP and identify the links to other plans, both within the AOR (or functional area) and with those of other CCMDs. The result of an IPR should include an endorsement of the planning to date or acknowledgement of friction points and guidance to shape continued planning. All four APEX operational activities (situational awareness, planning, execution, and assessment) continue in a complementary and iterative process. CJCSI 3141.01, *Management and Review of Joint Strategic Capabilities Plan (JSCP)-Tasked Plans,* provides further details on the IPR process.

(2) The JFC guides plan development by issuing a PLANORD or similar planning directive to coordinate the activities of the commands and agencies involved. A number of activities are associated with plan development, as Figure V-17 shows. These planning activities typically will be accomplished in a parallel, collaborative, and iterative fashion rather than sequentially, depending largely on the planning time available. The same flexibility displayed in COA development is seen here again, as planners discover and eliminate shortfalls and conflicts within their command and with the other CCMDs.

(3) The CJCS APEX family of documents referenced in CJCS Guide 3130, *Adaptive Planning and Execution Overview and Policy Framework*, provides policy, procedures, and guidance on these activities for organizations required to prepare a plan or order. These are typical types of activities that supported and supporting commands and Services accomplish collaboratively as they plan for joint operations.

(a) **Application of Forces and Capabilities**

1. When planning forces and capabilities, the commander is constrained by the total quantity of forces in the force apportionment tables. If additional resources are deemed necessary to reduce risk, CJCS approval is required. The supported commander should address the additional force requirement as early as possible in the IPR process, justify the requirement, and identify the risk associated if the forces are not made available. Risk assessments will include results using both apportioned capabilities and augmentation capabilities.

2. The supported commander should designate the **main effort** and **supporting efforts** as soon as possible and identify interdependent missions (especially subsequent tasks dependent on the successful completion of earlier tasks). This action is necessary for economy of effort. The main effort is based on the supported JFC's prioritized objectives. It identifies where the supported JFC will concentrate capabilities or prioritize efforts to achieve specific objectives. Designation of the main effort can be addressed in geographical (area) or functional terms. **Area tasks and responsibilities** focus on a specific area to control or conduct operations. An example is the assignment of

---

Plan Development Activities

- Force planning

- Support planning

- Nuclear strike planning

- Deployment and redeployment planning

- Shortfall identification

- Feasibility analysis

- Refinement

- Documentation

- Plan review and approval

- Supporting plan development

Figure V-17. Plan Development Activities

areas of operations for Army forces and Marine Corps forces operating in the same JOA. **Functional tasks and responsibilities** focus on the performance of continuing efforts that involve the forces of two or more Military Departments operating in the same OA or where there is a need to accomplish a distinct aspect of the assigned mission. An example is the designation of the maritime component commander as the joint force air component commander when the Navy component commander has the preponderance of the air assets and the ability to effectively plan, task, and control joint air operations. In either case, designating the main effort will establish where or how the JFC concentrates friendly forces and assets and/or prioritizes effort to attain an objective of an operation or campaign, or establish conditions that enable future operations that best support achieving subsequent objectives.

<u>3.</u> Designating a main effort facilitates the synchronized and integrated employment of the joint force by identifying priority missions when resources are limited while preserving the initiative of subordinate commanders. After the main effort is identified, joint force and component planners determine those tasks essential to accomplishing objectives. The supported JFC assigns these tasks to subordinate commanders along with the capabilities and support necessary to accomplish them. As such, the CONOPS must clearly specify the nature of the main effort.

<u>4.</u> The main effort can change during the course of the operation based on numerous factors, including changes in the OE, how the adversary reacts to friendly operations, including the successful accomplishment of previous missions or achievement of objectives. When the main effort changes, support priorities must change to ensure success. Both horizontal and vertical coordination within the joint force and with multinational and interagency partners is essential when shifting the main effort. Secondary efforts are important, but are ancillary to the main effort. They normally are planned to complement or enhance the success of the main effort (for example, by diverting enemy resources or setting conditions to enable the main operation). Only necessary secondary efforts, whose potential value offsets or exceeds the resources required, should be undertaken, because these efforts may divert resources from the main effort. Secondary efforts normally lack the operational depth of the main effort and have fewer forces and capabilities, smaller reserves, and more limited objectives.

(b) **Force Planning**

<u>1.</u> The primary purposes of force planning are to identify all forces needed to accomplish the CONOPS and effectively phase the forces into the OA. Force planning consists of determining the force requirements by operation phase, mission, mission priority, mission sequence, and operating area. It includes force requirements review, major force phasing, integration planning, and force list refinement. Force planning is the responsibility of the supported CCDR, supported by component commanders in coordination with the JS, JFPs, and FPs. Force planning begins early during plan development and focuses on applying the right force to the mission at the right time, while ensuring force visibility, force mobility, and adaptability. The commander determines force requirements and as necessary, develops a TPFDD letter of instruction specific to the OA and plans force modules to align and time-phase the forces in accordance

with the CONOPS. Proper force planning allows the ID of preferred forces to be selected for planning and included in the supported commander's CONOPS by operation phase, mission, and mission priority. Service components then collaboratively determine the specific sustainment capabilities required in accordance with the CONOPS. Upon direction to execute, the CCDR submits the refined RFF to the JS. CJCSM 3130.06, *(U) Global Force Management Allocation Policies and Procedures,* provides a detailed discussion of the GFM allocation process.

2. **Considerations.** The total force identified for supporting and supported plans is constrained by the quantity of forces identified in the force apportionment tables or otherwise prescribed force planning limitations. To support building a plan that can transition to execution, force requirements should be documented with the requisite data and information to support the GFM allocation process. This information informs the Services of the mission, tasks, purpose, priority, and specialized requirements for forces, as well as supports SecDef's decision making.

3. **Notional TPFDD Development.** Force requirements may be documented in a notional TPFDD and phased/sequenced during plan development to support the CONOPS. The notional TPFDD depicts force requirements and force flow. It is used to assess sourcing and transportation feasibility. When developed, the notional TPFDD will be entered into JOPES as the basis for this analysis. A notional TPFDD is used during planning and does not always contain execution sourced units.

4. **Preferred Force ID.** Developed as assumptions throughout planning, joint and component planners continue ID/refinement of specific units that satisfy the planned force requirements. The ID of preferred forces can be done informally at the command-level with component planners or it can leverage the JS J-35, JFPs, and FPs for a more detailed analysis and recommendation of force sourcing assumptions. Preferred forces are not sourced but provide critical assumptions essential for continued planning and sourcing and transportation feasibility analysis.

5. **Mobilization Planning.** Initial requirements for mobilization of Reserve Component force to include the scope and authorities should be identified early in planning. As preferred forces are refined, additional Reserve Component forces may also be identified. Timelines for mobilization should be developed in coordination with Service components and Service headquarters and incorporated into plan development.

6. **Non-DOD Capabilities.** Planner should document and refine non-DOD capabilities that are part of a plan CONOPS. Consideration should be made for interagency capabilities, nongovernmental capabilities including contracted support, and multinational capabilities. Most non-DOD capabilities are documented in a notional TPFDD to facilitate later analysis and potentially manage resources during execution and later in a TPFDD to document movement requirements.

7. **Rotational Requirements.** Rotational requirements are relevant if force rotations are envisioned to provide the requisite forces for long-term operations. When planning for operations that may be lengthy, consideration should be given to force

rotations. Typically force rotations are planned by Service headquarters in accordance with Service policy. Unit rotations should be timed so as to limit the impact on operations and rotational planning should consider JRSOI, turnover time, relief-in-place and transfer of authority, and time for the outbound unit to redeploy.

8. **Force Planning During Crisis.** Given the time constraints of a crisis, force planning may transition into execution sourcing vice preferred force ID. When force requirements are execution sourced, the TPFDD is populated with unit and movement data and becomes available for execution.

(c) **Support Planning.** Support planning is conducted concurrently with force planning to determine and sequence logistics and personnel support in accordance with the plan CONOPS. Support planning includes all core logistics functions: deployment and distribution, supply, maintenance, logistic services, OCS, health services, and engineering.

1. **Concept of Logistics Support.** Developed from the initial logistics staff estimate, the concept of logistics support is the foundation for logistics planning. This document provides an overview of the concept of support, priorities for movement of combat support forces and sustainment, identifies key logistics capabilities, and identifies metrics for assessing logistics effectiveness. CCMD planners must also consider the assignment of specific support responsibilities as follows:

a. **Directive Authority for Logistics.** CCDRs have directive authority for logistics and are authorized to provide authoritative direction to subordinate commands and forces necessary to carry out assigned missions. CCDRs may consider assigning responsibility for the planning, execution, and/or management of common support capabilities to a subordinate commander. This may streamline support operations for a plan, improve the support effectiveness, or eliminate duplication of effort among Service components.

b. **Lead Service.** A commander may choose to assign responsibility for planning and execution for one or more specific joint capability areas. This may be considered when a Service component provides the preponderance of support for a given core logistics function.

c. **Base Operating Support-Integrator.** When multiple Service components share a common base of operations, the supported commander may designate one of the Service components as the base operating support-integrator for that location to coordinate sustainment activities at that location.

d. **Partner Nation Support and HNS.** The JFC should also provide guidance for the use of partner nation support and HNS to meet support requirements. Partner nation support and HNS can provide support efficiencies but may not be appropriate or desired for all types of planning.

2. **Responsibilities.** Support planning is conducted by the Services, supporting commands, and CSAs in coordination with a supported CCMD's Service

components. The supported commander coordinates and synchronizes joint logistics to include communicating the support priorities to supporting commands and agencies. Service components and supporting organizations develop and refine their mission support, movement infrastructure, sustainment, and distribution plans in parallel with force planning.

For additional information on the joint deployment and distribution operation center and the GCC's options for assigning logistics responsibilities, see JP 4-0, Joint Logistics.

3. **Logistics supportability analyses (LSAs)** are conducted by supporting organizations to determine the logistics support they must provide, in accordance with resource informed planning guidance, and to determine the adequacy of resources needed to support mission execution. LSAs ensure logistics is phased to support the CONOPS; establishes logistics C2 authorities; and integrates support plans across the supporting commands, Service components, and agencies. LSAs are conducted by each supporting organization to the lowest level of detail needed to quantify the logistics requirements (national stock number level). These LSAs are then integrated by supporting organizations to coordinate roles and responsibilities, capabilities, and ensure all understand the sourcing of the support. A joint LSA is created and presented to the CCDR who confirms this support will provide the surge and sustainment needed to successfully execute and complete his mission. If there are gaps and shortfalls or high levels of risk that cannot be mitigated internally by supporting organization, the LSA provides the process for presenting issues to senior leaders for resolution.

4. **Transportation refinement** simulates the planned movement of resources that require lift, ensuring the plan is transportation feasible. The supported commander evaluates and adjusts the CONOPS to achieve end-to-end transportation feasibility if possible or requests additional resources if the level of risk is unacceptable. Transportation plans must be consistent and reconciled with plans and timelines required by providers of Service-unique combat and support aircraft to the supported CCDR. Planning must consider requirements of international law; commonly understood customs and practices; agreements or arrangements with foreign nations with which the US requires permission for overflight, access, and diplomatic clearance; en route infrastructure and destination port and airfield capacities. If significant changes are made to the CONOPS, it should be reassessed for transportation feasibility and refined to ensure it is acceptable.

(d) **Nuclear Strike Planning Options.** Commanders must assess the military as well as strategic impact a nuclear strike would have on conventional operations. Nuclear planning guidance is provided in Presidential policy documents and further clarified in DOD documents such as the GEF Annex B, and the Nuclear Supplement to the JSCP. Guidance issued to the CCDR is based on national-level considerations and supports the accomplishment of US objectives. United States Strategic Command (USSTRATCOM) is the lead organization for nuclear planning and coordination with appropriate allied commanders. The planning provided by USSTRATCOM ensures optimal integration of US nuclear and conventional forces prior to, during, and after conflict. USSTRATCOM uses this framework to develop detailed mission plans to be executed by the appropriate nuclear forces. USSTRATCOM coordinates with appropriate commanders to accomplish target deconfliction and preclusion analysis and ensures appropriate weapon yields, delivery methods, and safe delivery routing. Due to the strategic and diplomatic consequences associated with nuclear operations and plans, only the President has the authority to employ nuclear weapons.

(e) **Deployment and Redeployment Planning.** Deployment and redeployment planning is conducted on a continuous basis for all approved contingency plans and as required for specific crisis action plans. Planning for redeployment should be considered throughout the operation and is best accomplished in the same time-phased process in which deployment was accomplished. In all cases, mission requirements of a specific operation define the scope, duration, and scale of both deployment and redeployment operation planning. Unity of effort is paramount, since both deployment and redeployment operations involve numerous commands, agencies, and functional processes. Procedures and standards to attain and maintain visibility of personnel must be formulated. Because the ability to adapt to unforeseen conditions is essential, supported CCDRs must ensure their deployment plans for each contingency or crisis action plan support global force visibility requirements. When operations that may be lengthy are planned, consideration must be given to force rotations. Units must rotate without interrupting operations. Planning should consider JRSOI, turnover time, relief-in-place, transfer of authority, and time it takes for the outbound unit to redeploy. This information is vital for the FPs, JFPs, and JS J-35 to develop force rotations and order them in the GFMAP if the operation is executed.

1. **OE.** For a given plan, deployment planning decisions are based on the anticipated OE, which may be permissive, uncertain, or hostile. The anticipated OE dictates the deployment concept, which may require forcible entry operations. Normally, supported CCDRs, their subordinate commanders, and their Service components are responsible for providing detailed situation information, mission statements by operation phase, theater support parameters, strategic and operational lift allocations by phase (for both force movements and sustainment), HNS information and environmental standards, OCS aspects of the OE information, and pre-positioned equipment planning guidance.

2. **Deployment and Redeployment Concept.** Supported CCDRs must develop a deployment concept and identify specific predeployment standards necessary to meet mission requirements. Services and supporting CCDRs provide trained and mission-ready forces to the supported CCMD deployment concept and predeployment standard. Services recruit, organize, train, and equip interoperable forces. The Services' predeployment planning and coordination with the supporting CCMD must ensure predeployment standards specified by the supported CCDR are achieved, supporting personnel and forces arrive in the supported theater fully prepared to perform their mission, and deployment delays caused by duplication of predeployment efforts are eliminated. The Services and supporting CCDRs must ensure unit contingency plans are prepared, forces are tailored and echeloned, personnel and equipment movement plans are complete and accurate, command relationship and integration requirements are identified, mission-essential tasks are rehearsed, mission-specific training is conducted, force protection is planned and resourced, and both logistics and personnel services support sustainment requirements are identified. Careful and detailed planning makes certain that only required

personnel, equipment, and materiel deploy; unit training is exacting; missions are fully understood; deployment changes are minimized during execution; and the flow of personnel, equipment, and movement of materiel into theater aligns with the CONOPS. Supported CCDRs should also develop a redeployment CONOPS to identify how forces and materiel will either redeploy to home station or to support another JFC's operation. This redeployment CONOPS is especially relevant and useful if force rotations are envisioned to provide the requisite forces for a long-term operation. CCDRs may not have all planning factors to fully develop this CONOPS, but by using the best available information for redeployment requirements, timelines, and priorities, the efficiency and effectiveness of redeployment operations may be greatly improved. Topics addressed in this early stage of a redeployment CONOPS may include a proposed sequence for redeployment of units, individuals, materiel, and contract closeout and changes to the contractor management plan. Responsibilities and priorities for recovery, reconstitution, and return to home station may also be addressed along with transition requirements during mission handover. As a campaign or operation moves through the different operational plan phases, the CCDR will be able to develop and issue a redeployment order based on a refined redeployment CONOPS. Effective redeployment operations are essential to ensure supporting Services and rotational forces have sufficient time to fully source and prepare for the next rotation.

For additional information on deployment and redeployment planning, see JP 3-35, Deployment and Redeployment Operations.

   3. **Movement Planning.**   Movement planning is the collaborative integration of movement activities and requirements for transportation support. Forces may be planned for movement either by self-deploying or the use of organic lift and non-organic, common-user, strategic lift resources identified for planning. Competing requirements for limited strategic lift resources, support facilities, and intra-theater transportation assets will be considered by the supported commander in terms of impact on mission accomplishment. If additional resources are required, the supported commander will identify the requirements and rationale for those resources.

   a. **TPFDD Letter of Instruction.**   Commanders will often publish revised TPFDD development guidance articulating the commander's deployment and redeployment priorities. Planners then develop a final refinement of the plan's TPFDD in accordance with this revised guidance.

   b. **TPFDD Development.**   In order to conduct movement planning, the TPFDD must have specific unit assumptions identified for the required forces and equipment. These specific unit assumptions can be identified through preferred forces or contingency sourcing. Planners should leverage the expertise of the JS J-35, JFPs, and FPs in the development of specific unit assumptions.

   c. **Coordination with USTRANSCOM.**   The supported commander and USTRANSCOM coordinate to resolve transportation feasibility issues impacting intertheater and intratheater movement and sustainment delivery. USTRANSCOM and other transportation providers identify air, land, and sea transportation resources to support the CONOPS. These resources may include intertheater transportation, GCC-controlled theater transportation, and transportation organic to the subordinate commands. USTRANSCOM and other transportation providers develop transportation schedules for movement requirements identified by the supported commander. A transportation schedule does not imply the supported commander's plan is transportation feasible; rather, the schedule provides the most effective and realistic use of available transportation resources in relation to the plan.

    <u>d.</u> **JRSOI Planning.** Following the development of movement infrastructure concepts, the supported commander's planning team develops the air and sea reception plan, staging plan, and completed JRSOI plan. The requirements to conduct JRSOI may precipitate additional force requirements and cause iterative changes to force planning. JRSOI constraints (e.g., port clearance, intratheater movement capacity, staging base limitations) imposed on strategic movement must be considered in JRSOI planning and reflected in the TPFDD and TPFDL.

  d. **Shortfall ID.** Along with hazard and threat analysis, shortfall ID is conducted throughout the plan development process. The supported commander continuously identifies limiting factors, capability shortfalls, and associated risks as plan development progresses. Where possible, the supported commander resolves the shortfalls and required controls and countermeasures through planning adjustments and coordination with supporting and subordinate commanders. If the shortfalls and necessary controls and countermeasures cannot be reconciled or the resources provided are inadequate to perform the assigned task, the supported commander reports these limiting factors and assessment of the associated risk to the CJCS. The CJCS and the JCS consider shortfalls and limiting factors reported by the supported commander and coordinate resolution. However, the completion of plan development is not delayed pending the resolution of shortfalls. If shortfalls cannot be resolved within the prescribed timeframe, the completed plan will include a consolidated summary, including the impact of unresolved shortfalls and associated risks.

  e. **Feasibility Analysis.** This step in plan or order development is similar to determining the feasibility of a COA, except it typically does not involve simulation-based wargaming. The focus in this step is on ensuring the assigned mission can be accomplished using available resources within the time contemplated by the plan. The results of force planning, support planning, deployment and redeployment planning, and shortfall ID will affect feasibility. The primary factors to consider are the capacity of lift and throughput constraints of transit points and JRSOI infrastructure that can support the plan. The primary factors analyzed for feasibility include forces, resources, and transportation.

    (1) **Forces.** The supported commander, in coordination with the JS J-35, FPs, JFPs, and Military Departments, should determine the feasibility of sourcing the plans required forces. For all planning, the sourcing feasibility analysis should consider the total force requirements of supported and supporting plans. Force requirements should be documented in the plan's TPFDD for subsequent transportation feasibility and enable GFM allocation should the plan transition to execution.

(2) **Sustainment Resources.** The supported commander, in coordination with Military Departments and CSAs, should determine the feasibility of providing the resources required to execute the plan. Supporting organizations must provide subject matter experts to identify sustainment requirements and gaps. As with forces, analysis of sustainment requirements should consider the total requirements of supported and supporting plans. Sustainment requirements that require movement should be documented in the plan's TPFDD for inclusion in overall transportation feasibility.

(3) **Transportation.** The supported commander, in coordination with the JS and USTRANSCOM, determine the transportation feasibility of a plan. Transportation feasibility requires the assumed sourcing of forces through preferred force ID or contingency sourcing to create a notional TPFDD. The plan's notional TPFDD reflects these sourcing assumptions and identifies transportation requirements of forces and resources for this analysis.

f. **Documentation.** When the TPFDD is complete and end-to-end transportation feasibility has been achieved and is acceptable to the supported CCDR, the supported CCDR completes the documentation of the plan or OPORD and coordinates access with respective JPEC stakeholders to the TPFDD as appropriate.

g. **Movement Plan Review and Approval.** When the plan or OPORD is complete, JS J-5 coordinates with the JPEC for review. The JPEC reviews the plan or OPORD and provides the results of the review to the supported CCDR and CJCS. The CJCS reviews and provides recommendations to SecDef, if necessary. The JCS provides a copy of the plan to OSD to facilitate their parallel review of the plan and to inform USD(P)'s recommendation of approval/disapproval to SecDef. After the CJCS's and USD(P)'s review, SecDef or the President will review, approve, or modify the plan. The President or SecDef is the final approval authority for OPORDs, depending upon the subject matter.

See CJCSI 3141.01, Management and Review of Joint Strategic Capabilities Plan (JSCP)-Tasked Plans, *for more information on plan review and approval.*

h. **Transition.** Transition is an orderly turnover of a plan or order as it is passed to those tasked with execution of the operation. It provides information, direction, and guidance relative to the plan or order that will help to facilitate situational awareness. Additionally, it provides an understanding of the rationale for key decisions necessary to ensure there is a coherent shift from planning to execution. These factors coupled together are intended to maintain the intent of the CONOPS, promote unity of effort, and generate tempo. Successful transition ensures those charged with executing an order have a full understanding of the plan. Regardless of the level of command, such a transition ensures those who execute the order understand the commander's intent and CONOPS. Transition may be internal or external in the form of briefs or drills. Internally, transition occurs between future plans and future/current operations. Externally, transition occurs between the commander and subordinate commands.

(1) **Transition Brief.** At higher levels of command, transition may include a formal transition brief to subordinate or adjacent commanders and to the staff supervising

execution of the order. At lower levels, it might be less formal. The transition brief provides an overview of the mission, commander's intent, task organization, and enemy and friendly situation. It is given to ensure all actions necessary to implement the order are known and understood by those executing the order. The brief may include items from the order or plan such as:

(a) Higher headquarters' mission and commander's intent.

(b) Mission.

(c) Commander's intent.

(d) CCIRs.

(e) Task organization.

(f) Situation (friendly and enemy).

(g) CONOPS.

(h) Execution (including branches and potential sequels).

(i) Planning support tools (such as a synchronization matrix).

(2) **Confirmation Brief.** A confirmation brief is given by a subordinate commander after receiving the order or plan. Subordinate commanders brief the higher commander on their understanding of commander's intent, their specific tasks and purpose, and the relationship between their unit's missions and the other units in the operation. The confirmation brief allows the higher commander to identify potential gaps in the plan, as well as discrepancies with subordinate plans. It also gives the commander insights into how subordinate commanders intend to accomplish their missions.

(3) **Transition Drills.** Transition drills increase the situational awareness of subordinate commanders and the staff and instill confidence and familiarity with the plan. Sand tables, map exercises, and rehearsals are examples of transition drills.

(4) **Plan Implementation.** Military plans and orders should be prepared to facilitate implementation and transition to execution. For a plan to be implemented, the following products and activities must occur:

(a) Confirm assumptions. Analyze the current OE and establish as fact any assumptions made during plan development.

(b) Model the TPFDD to confirm the sourcing and transportation feasibility assessment. Validate that force and mobility resources used during plan development are currently available.

(c) Establish execution timings. Set timelines to initiate operations to allow synchronization of execution.

(d) Confirm authorities for execution. Request and receive the President or SecDef authority to conduct military operations.

(e) Conduct execution sourcing from assigned forces or request allocation of required forces. Identify specific units to allocate against CCDR force requirements based upon current conditions and risk evaluation. Develop new assumptions, if required.

(f) Issues necessary orders for execution. The CJCS issues orders implementing the directions of the President or SecDef to conduct military operations. CCDRs subsequently issue their own orders directing the activities of subordinate commanders.

CHAPTER VI
OPERATION ASSESSMENT

"However beautiful the strategy, you should occasionally look at the results."

Winston Churchill, Prime Minister of Great Britain

SECTION A. GENERAL

1. Overview

a. Operation assessments are an integral part of planning and execution of any operation, fulfilling the requirement to identify and analyze changes in the OE and to determine the progress of the operation. Assessments involve the entire staff and other sources such as higher and subordinate headquarters, interagency and multinational partners, and other stakeholders. They provide perspective, insight, and the opportunity to correct, adapt, and refine planning and execution to make military operations more effective. Operation assessment applies to all levels of warfare and during all military operations.

KEY TERM

Assessment: A continuous activity that supports decision making by ascertaining progress toward accomplishing a task, creating an effect, achieving an objective, or attaining an end state for the purpose of developing, adapting, and refining plans and for making campaigns and operations more effective.

---

b. Commanders maintain a personal sense of the progress of the operation or campaign, shaped by conversations with senior and subordinate commanders, key leader engagements (KLEs), and battlefield circulation. Operation assessment complements the commander's awareness by methodically identifying changes in the OE, identifying and analyzing risks and opportunities, and formally providing recommendations to improve progress towards mission accomplishment. Assessment should be integrated into the organization's planning (beginning in the plan initiation step) and operations battle rhythm to best support the commander's decision cycle.

c. The starting point for operation assessment activities coincides with the initiation of joint planning. Integrating assessments into the planning cycle helps the commander ensure the operational approach remains feasible and acceptable in the context of higher policy, guidance, and orders. This integrated approach optimizes the feedback senior leadership needs to appropriately refine, adapt, or terminate planning and execution to be effective in the OE.

d. CCMDs, subordinate Service, joint functional components, and JTFs devote significant effort and resources to plan and execute operations. They apply appropriate rigor to determine whether an operation is being effectively planned and executed as needed to accomplish specified objectives and end states. Assessment complements that rigor by analyzing the OE objectively and comprehensively to estimate the effectiveness of planned tasks and measure the effectiveness of completed tasks with respect to desired conditions in the OE.

e. **Background**

(1) TCPs and country-specific security cooperation sections/country plans are continuously in some stage of implementation. Accordingly, during implementation CCMD planners should annually extend their planning horizon into the future year. The simultaneity of planning for the future while implementing a plan requires a CCMD to continually assess its implementation in order to appropriately revise, adapt, or terminate elements of the evolving (future) plan. This synergism makes operation assessment a prerequisite to plan adaptation. Operation assessment is thus fundamental to revising implementation documents ahead of resource allocation processes.

(a) Events can arise external to the CCMD's control that affect both plan execution and future planning. Some of these events can impede achievement of one or more objectives while others may present opportunities to advance the plan more rapidly than anticipated.

(b) External events generally fall into two categories. The first are those that change the strategic or OE in which a CCMD implements a plan (typically a J-2 focus). The second category involves those events that change the resource picture with respect to funding, forces, and time available (typically a force structure, resource, and assessment directorate of a joint staff [J-8] focus). This document treats these two types of external events as separate considerations because they can influence plan implementation independent of each other.

(2) Throughout campaign planning and execution, the CCDR and staff continually observe the OE and assess the efficacy of the campaign plan. Operation assessment at the CCMD level is often referred to as theater or global campaign assessment or, generically, as campaign assessment. Because campaigns are conducted in a complex and dynamic environment, commands must be able to detect, analyze, and adapt to changes in the OE during execution. Planners review the guidance, their understanding of the OE, the campaign objectives, and the decisions that underpinned the original operational approach to refine or adapt the plan, the approach, or the guidance.

(3) In addition to the command's internal assessment efforts, analysis and assessment of the strategic and OEs by interagency partners is available to the CCMD. OSD and the JS can assist in obtaining these inputs. Promote Cooperation events enable interagency partners' insights on environmental changes to be shared with the CCMDs.

For more information on Promote Cooperation events, see CJCSM 3130.01, Campaign Planning Procedures and Responsibilities.

(4) The overall purpose of operation assessment is to provide recommendations to make operations more effective. As it relates to campaigns, where strategic objectives

frame the CCMD's mission, operation assessment helps the CCDR and supporting organizations refine or adapt the campaign plan and supporting plans to achieve the campaign objectives or, with SecDef, to adapt the GEF-directed objectives to changes in the strategic objectives and OEs.

(5) The assessment process serves as part of the CCMD's feedback mechanism throughout campaign planning and execution. It also feeds external requirements such as the CCDR's inputs to the CJCS AJA. Assessment analysis and products should identify where the CCMD's ways and means are sufficient to attain their ends, where they are not and why not, and support recommendations to adapt or modify the campaign plan or its components. The analyses might provide insight into basic questions such as:

(a) Are the objectives (strategic and intermediate) achievable given changes in the OE and emerging diplomatic/political issues?

(b) Is the current plan still suitable to achieve the objectives?

(c) Do changes in the OE impose additional risks or provide additional opportunities to the command?

(d) To what degree are the resources employed making a difference in the OE?

(6) Campaign assessment analyses and products should provide the CCDR and staff with sufficient information to make, or recommend, necessary adjustments to plans, policy, resources, and/or authorities in the next cycle of planning to make operations more effective. Assessment can be used to inform OSD and CJCS reporting requirements as mandated by the GEF and other strategic planning documents.

(a) Campaign assessment activities should facilitate the CCDR's input to DOD on the capabilities and authorities needed to accomplish the missions in the CCMD's contingency plans over the CCDR's strategic planning horizon. The campaign assessment should take into account expected changes in threats and the strategic and OEs.

(b) Campaign assessment analyses and products should also help the CCDR request additional resources or to recommend re-allocating available resources to desired priorities. Assessment analyses and products likewise inform SecDef and senior leaders' resourcing decisions across all CCMDs and DOD requests to Congress to add or reallocate resources through the FYDP.

f. **Campaign Assessments**

(1) Campaign assessments determine whether progress towards achieving CCMD campaign objectives is being made by evaluating whether progress towards intermediate objectives is being made. Intermediate objectives are desired conditions in the OE the CCDR views as critical for successfully executing the campaign plan and achieving CCMD campaign objectives. Essentially, intermediate objectives are multiple objectives that are between initiation of the campaign and achievement of campaign

objectives. Accordingly, at the strategic assessment level, intermediate objectives are criteria used to observe and measure progress toward campaign desired conditions and evaluate why the current status of progress exists.

(a) Functional campaign assessments assist the FCCs in evaluating progress toward or regression from achieving their global functional objectives. FCCs provide unique support to GCCs in their respective specialties and are required to assess progress towards their intermediate objectives in support of their global functional objectives or DOD-wide activities.

(b) The CJCS is responsible for aggregating CCDRs' campaign plan assessments and setting assessment standards for functional objectives and DOD-wide activities. CCMD FCP and DOD-wide activities campaign plan assessments will be compiled into this assessment framework to inform an integrated evaluation of global progress against geographic and functional objectives. Planners responsible for developing global campaign plans will collaborate with CCDRs on common LOEs and intermediate objectives that affect functional objectives (i.e., distribution or DOD-wide activities).

(2) The JSCP, GEF, and other strategic guidance provide CCMDs with strategic objectives. CCMDs translate and refine those long-range objectives into near-term (achievable in 2-5 years) intermediate objectives. Intermediate objectives represent unique military contributions to the achievement of strategic objectives. In some cases, the CCMD's actions alone may not achieve strategic objectives; additional intermediate objectives may be required to achieve them. Consequently, other instruments of national power may be required, with the CCMD operating in a supported or supporting role.

(3) The basic process for campaign assessment is similar to that used for contingency and crisis applications but the scale and scope are generally much larger. While operational level activities such as a JTF typically focus on a single end state with multiple desired conditions, the campaign plan must integrate products from a larger range of strategic objectives, each encompassing its own set of intermediate objectives and desired conditions, subordinate operations, and subordinate plans (i.e., country-specific security cooperation sections/country plans, contingency plans not in execution, on-going operations, directed missions) (see Figure VI-1).

(4) One common method to establish more manageable campaign plans is for CCMDs to establish LOEs with associated intermediate objectives for each campaign objective. This method allows the CCMD to simultaneously assess each LOE and then assess the overall effort using products from the LOE assessments. The following discussion uses several boards, cells, and working groups. The names merely provide context for the process and are not intended to be a requirement for organizations to follow (see Figure VI-2).

(5) The assessment needs to nest with and support the campaign and national objectives and cannot rely on accomplishment of specific tasks. Commanders and staffs should make certain the established intermediate objectives will change the OE in the manner desired.

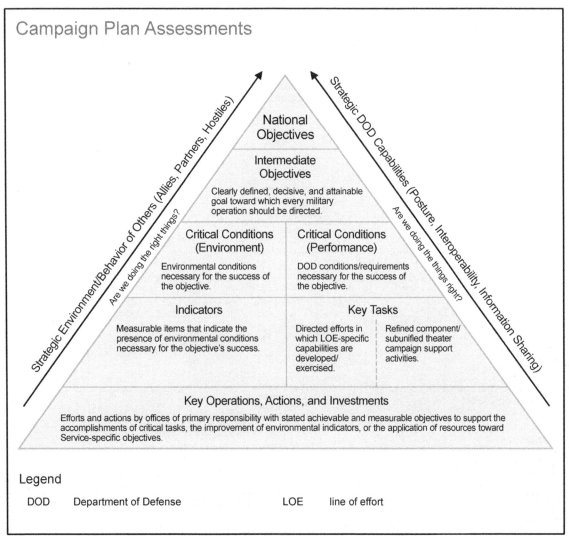

Figure VI-1. Campaign Plan Assessments

(a) **LOE Assessment**

1. **Leads.** LOE leads should guide the development and assessment of LOE intermediate objectives, critical conditions, indicators, tasks, and associated metrics and recommendations through the LOE working groups.

2. **Output.** The LOE assessment produces updated findings, insights, and recommendations by LOE. These are consolidated for presentation and validation during the strategic assessment working group (SAWG).

(b) **SAWG**

1. **Leads.** Designated lead (typically from a J-3, J-5, or J-8 element) chairs this O-6 level review working group. LOE assessors and leads brief their sub-campaign assessments, findings, insights, and recommendations to this group.

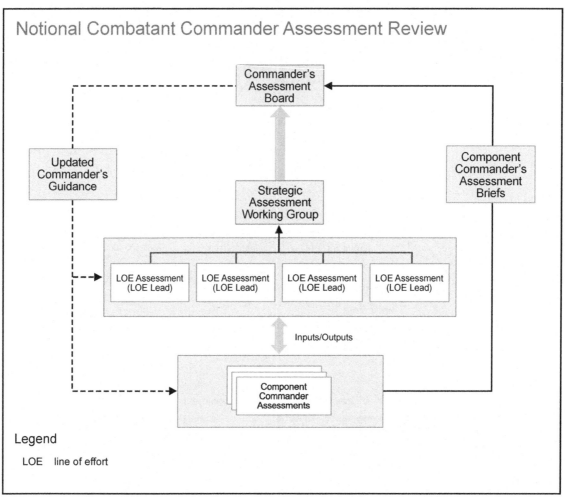

Figure VI-2. Notional Combatant Commander Assessment Review

2. **Output.** The SAWG produces an assessment brief and recommendations for presentation and approval during the commander's assessment board (CAB).

(c) **CAB**

1. **Leads.** CCDRs chair this board. LOE leads present a consolidated assessment brief with SAWG-validated, command-level recommendations for the commander's decision. As a note, this board may occur as part of the commander's council or the commander's update brief.

2. **Outputs.** The CAB validates recommendations for staff action and higher level coordination and produces refined commander's guidance.

(6) **Component Command Assessment.** If required by the CCDR, component and subordinate commands will provide an annual assessment briefing to CCDR detailing their progress toward key LOE objectives and conduct of key operations and activities.

2. The Purpose of Operation Assessment in Joint Operations

a. Operation assessments help the commander and staff determine progress toward mission accomplishment. Assessment results enhance the commander's decision making enable more effective operations and help the commander and the staff to keep pace with a constantly evolving OE. A secondary purpose is to inform senior civil-military leadership dialogue to support geopolitical and resource decision making throughout planning and execution.

b. Integrating assessment during planning and execution can help commanders and staffs to:

(1) Develop mission success criteria.

(2) Compare observed OE conditions to desired objectives and/or end state conditions.

(3) Determine validity of key planning facts and assumptions.

(4) Determine whether or not the desired effects have been created and whether the objectives are being achieved.

(5) During execution, determine the effectiveness of allocated resources against specific task and mission performance and effects, and test the validity of intermediate objectives.

(6) Determine whether an increase, decrease, or change to resources is required.

(7) Identify the risks and barriers to mission accomplishment.

(8) Identify opportunities to accelerate mission accomplishment.

3. Tenets of Operation Assessment

The following tenets should guide the commander and the staff throughout assessment:

a. **Commander Centricity.** The commander's involvement in operation assessment is essential. The assessment plan should focus on the information and intelligence that directly support the commander's decision making.

b. **Subordinate Commander Involvement.** Assessments are more effective when used to support conversations between commanders at different echelons. Operation assessments link echelons of command by identifying the activities and impacts critical to success and sharing the assessment methods used to shape operational decisions. A common understanding of operational priorities allows subordinate commanders to directly communicate their most relevant information.

c. **Integration.** Staff integration is crucial to planning and executing effective assessments. Operation assessment is the responsibility of commanders, planners, and operators at every level and not the sole work of an individual advisor, committee, or assessment entity. It is nested within the planning process and integrates roles across the staff. Properly structured, operation assessments enable the staff to examine and understand how actions are related. Integrating perspectives from across the staff should minimize errors that arise from limited focus (i.e., duplication of effort, incorrect ID of causes, or insufficient information to prioritize issues by level of impact).

d. **Integration into the Planning Process and Battle Rhythm.** To deliver information at the right time, the operation assessment should be synchronized with the commander's decision cycle. The assessment planning steps occur concurrently with the steps of JPP. The resulting assessment plan should support the command's battle rhythm.

e. **Integration of External Sources of Information.** Operation assessment should allow the commander and staff to integrate information that updates the understanding of the OE in order to plan more effective operations. To get a more complete understanding of the OE, it is important to share relevant information with the HN, interagency, multinational, private sector, and nongovernmental partners. For aspects of the OPLAN for which nonmilitary influence has high impact or is not well understood, input from these sources is critical to refine understanding of the OE and to reduce risk.

f. **Credibility and Transparency.** Assessment reports should cite all sources of information used to build the report. The staff should use methods that are appropriate to the environment and to the task of assessing a complex operation. As much as possible, sources and assessment results should be unbiased. All methods used, and limitations in the collection of information and any assumptions used to link evidence to conclusions, should be clearly described in the assessment report.

g. **Continuous Operation Assessment.** While an operation assessment product may be developed on a specific schedule, assessment is continuous in any operation. The information collected and analyzed can be used to inform planning, execution, and assessment of operations.

4. **Commander and Staff Involvement**

a. The commander's requirements for decision making should focus the assessment plan and activity. Assessment is a key component of the commander's decision cycle, helping to determine the results of operations, activities, and investments in the context of the overall mission objectives and providing recommendations for the refinement of plans and orders. If assessment products and analyses do not provide the commander with answers to specific questions pertaining to recommended actions to improve operational progress, acting on opportunities, or mitigating risks, they do not provide value.

b. Commanders establish priorities for assessment through their planning guidance, CCIRs, and decision points. Commanders tell their staff and subordinate commanders what they need, when (how often) they need it, and how they wish to receive it. Commanders also give their staffs guidance on where to focus limited collection and analytical resources.

(1) Commanders and staff must balance collection and monitoring efforts between "what they can know" versus "what they need to know." The collecting and monitoring effort should reject the tendencies to: measure things simply because they are measurable, demand measures where valid data does not exist, or ignore something pertinent because it is hard to measure. Understanding the difference may also help commanders and their staffs avoid burdening subordinates with overly detailed assessment and collection tasks.

(2) Commanders should leverage staff and subordinate commander assessments, personal observation of the OA, discussions with stakeholders, and experience and instincts to formulate their own assessment.

(3) Commanders should regard a plan which does not include assessment considerations and guidance as incomplete.

c. Assessment informs and strengthens the commander's understanding of the OE. Effective staffs leverage and integrate planning and operations processes and existing reporting mechanisms whenever possible to enable synchronized assessments without adding significant additional requirements to personnel and subordinate units.

d. Significant challenges that staffs must often overcome to enable an effective operation assessment activity include:

(1) **Integrating assessment into planning and execution from the outset.** The ongoing activities of situational awareness and assessment shape ongoing planning and execution and influence the overall decision cycle of the commander. The most successful staffs are those that routinely integrate and implement assessment activity at the onset of the planning process. Concurrently considering operational assessment during planning supports the development of well-written objectives. Failing to consider how to assess an operation during planning can lead to objectives that do not lend themselves to measurement towards achievement of the objectives and tasks directed to staff and subordinate commands that are not tied to or support operational objectives.

(2) **Failing to conduct adequate analysis before acting.** The assessment process, which includes detailed JIPOE products, is designed to improve the understanding of the OE, including understanding of the causal links between friendly operations, activities, and investments and changes in the OE, creating conditions favorable to mission accomplishment, and identifying actionable opportunities and risk mitigation measures to improve the likelihood of mission success. Close coordination between the assessment staff and intelligence personnel conducting JIPOE will also support brainstorming effective requests for information for baseline data. Adequate analytic rigor is required to address complex issues to portray recommendations accurately. The staff should recognize the entire breadth of assessment contributing to the assessment operational activity of the command. The staff should consider leveraging already existing assessments and how the

assessment of a specific plan contributes to the overall decision cycle of the command's higher headquarters. Recommendations provided by the assessment process provide insight to the commander and staff, enabling adjustments to current operations, activities, and investments and identifying planning refinement and adaptation efforts to enhance operational effectiveness.

(3) **Ensuring assessment keeps pace with a commander's appreciation of the OE.** The commander's understanding of the OE is driven by continual interaction with subordinate commanders, KLEs, stakeholders, and battle space circulation. Conversely, most staffs must rely upon information provided by operational and intelligence reporting, usually within a set battle rhythm, and requiring consolidation, analysis, and some level of cross-staff vetting, often in the form of an assessment working group. Therefore, formal assessment reports and briefings are often delivered behind the pace of the operation. Further, senior staff and commanders will not wait for formal reports to act when necessary to adjust the operation. Staffs should leverage the collection management process, effectively calibrate assessment activity to the pace of operations, and recalibrate assessment requirements as the operation progresses in order to keep pace with and contribute meaningfully to the commander's understanding of the OE.

(4) **Ensuring recommendations facilitate the commander's decision making.** The staff must consider what kinds of decisions the commander will have to make in order to achieve objectives and attain the end state. Decisions include both internal and external action. As such, recommendations developed during the assessment process should not be limited to only those resources and authorities over which the commander has control. Optimally, assessment recommendations should facilitate the commander's ability to provide guidance and directions to subordinates, request additional support from supporting organizations, and recommend additional diplomatic, informational, military, or economic actions to interagency and multinational partners.

(5) **Resolving cross-organization resistance to assessment process requirements.** All staff directorates should be aware of the importance of operations assessment to the commander as incomplete or missing data could lead to an inaccurate assessment and faulty decisions. Operation assessment is a cross-command process, and developing ownership in the process and briefings (for example, where insights and recommendations are presented to the commander by LOE working groups at the O-6 level) stimulates broader interest and quality.

(6) **Integrating joint force component activities and efforts into the campaign assessment process.** In most CCMDs, joint force components own most of the resources that operationalize the campaign plan. They will be focused on their own component support plan and Title 10, USC, activities, so it may require more effort by the CCDR's campaign assessment process to make certain component operations and activities can be developed and focused to attain the CCDR's objectives.

(7) **Lack of advocacy or commander disinterest.** Senior staff needs to ensure the commander appreciates the value of assessment and strives to meet the assessment needs.

5. Staff Organization for Operation Assessment

a. Cross-functional staff representation is required to effectively analyze progress toward achieving objectives. This provides the assessment activity with varied perspectives and broad expertise that are necessary for the assessment's credibility and rigor.

b. Roles and responsibilities for the assessment team is a key consideration. The ability to work across the staff will impact the quality and relevance of assessment efforts. The commander or COS should identify the director or staff entity responsible for the collective assessment effort in order to synchronize activities, achieve unity of effort, avoid duplication of effort, and clarify assessment roles and responsibilities across the staff. The assessment activity should be routine and not ad hoc. The responsible director or staff entity should have the authority to integrate and synchronize the staff when conducting the assessment process. The COS should play a pivotal role in staff synchronization for operation assessments, as the COS typically leads the command's operational cycle. Within typical staff organizations there are three basic locations where the responsible element could reside:

(1) **Special Staff Section.** In this approach, the assessment element reports directly to the commander, via the COS or deputy commander. Advantages of this approach may include increased access to the commander and visibility on decision making requirements, as well as an increased ability to make recommendations to the commander as part of the assessment process. Disadvantages may include being isolated from the other staff sections and not having access to the information being collected and monitored across the staff.

(2) **Separate Staff Section.** In this approach, the assessment element is its own staff section, akin to plans, operations, intelligence, logistics, and communications. The advantage of this approach is that it legitimizes assessment as a major staff activity equivalent with the other staff functions and allows the assessment team to participate in staff coordination and activities as co-equals with the other staff sections. A disadvantage to this approach is that it has the potential to create stove-piped assessment efforts without full collaboration for a whole of staff assessment.

(3) **Integrated in Another Staff Section.** In this approach, the assessment element is typically integrated into the operations or plans sections, and the assessment chief reports to the plans chief or the operations chief. The advantage of this approach is that it tends to create close ties between the assessment team and either the plans or operations teams, but a significant disadvantage is that this approach limits the access of the assessment team to the commander and other elements of the staff and typically introduces another layer of review (and potential bias) of the assessment team's products.

SECTION B. CONDUCTING OPERATION ASSESSMENT

6. General

The assessment process is continuous. Throughout JPP, assessment provides support to and is supported by operational design and operational art. The assessment process complements and is concurrent with JPP in developing specific and measurable task-based end states, objectives, and effects during operational design. These help the staff identify the information and intelligence requirements (including CCIRs). During execution, assessment provides information on progress toward creating effects, achieving objectives, and attaining desired end states. Assessment reports are based on continuous situational awareness and OE analysis from internal and external sources and address changes in the OE and their proximate causes, opportunities to exploit and risks to mitigate, and recommendations to inform decision making throughout planning and execution.

7. Operation Assessment Process

There is no single way to conduct assessment. Every mission and OE has its own unique challenges, making every assessment unique. The following steps can help guide the development of an effective assessment plan and assessment performance during execution. Assessment steps provide an orderly, analytical process to help organizations understand the underpinnings of desirable or undesirable action or behavior. Organizations should consider these steps as necessary to fit their needs. Figure VI-3 provides an overview of the assessment process, which is further explained in subsequent paragraphs.

a. **Step 1—Develop the Operation Assessment Approach**

(1) Operation assessment begins during the initiation step of JPP when the command identifies possible operational approaches and their associated objectives, tasks, effects, and desired conditions in the OE. Concurrently, the staff begins to develop the operation assessment approach by identifying and integrating the appropriate assessment plan framework and structure needed to assess planning and execution effectiveness. The assessment approach identifies the specific information needed to monitor and analyze effects and conditions associated with achieving operation or campaign objectives. The assessment approach becomes the framework for the assessment plan and will continue to mature through plan development, refinement, adaptation, and execution in order to understand the OE and measure whether anticipated and executed operations are having the desired impact on the OE (see Figure VI-4). In short, the command tries to answer the following questions: "How will we know we are creating the desired effects," "Are we achieving the objectives," "What information do we need," and "Who is best postured to provide that information."

(2) The first step of the assessment approach aligns to all JPP steps, as assessment should complement and be concurrent with the planning effort. A common error is not considering assessment until the plan is completed. Upon receipt of a new mission or

Operation Assessment Steps

Activity	Primarily in Planning or Execution	Personnel Involved	Input	Associated Staff Activity	Output

Develop Assessment Approach	Planning
	• Operational Design
	• JPP Steps 1-6
		• Commander
		• Planners
		• Primary staff
		• Special staff
		• Assessment element
			Strategic guidance
			CIPG
			Description of OE
			Problem to be solved
			Operational approach
			Commander's intent (purpose, end state, risk)
				• Conduct JIPOE
				• Develop operational approach
				• Support development and refinement of end states, objectives, effects, and tasks
				• Conduct joint planning (JPP and operational design)
				• Determine and develop how to assess tasks, effects, objectives, and end state progress for each course of action:
				• Identify indicators
					Assessment approach which includes:
					  assessment framework and construct
					  Specific outcomes (end state, objectives, effects)
					  Commander's estimate/CONOPS (from JPP)

Develop Assessment Plan	Planning
	• JPP Step 7
		• Commander
		• Planners
		• Primary staff
		• Special staff
		• Assessment element
		• Operations planners
		• Intelligence planners
		• Subordinate commanders
		• Interagency and multinational partners
		• Others, as required
			Assessment approach which includes:
			assessment framework and construct
			Specific outcomes (end state, objectives, effects)
			Commander's estimate/ CONOPS (from JPP)
				• Document assessment framework and construct
				  ○ Finalize the data collection plan
				  ○ Coordinate and assign responsibilities for monitoring, collection, and analysis
				  ○ Identify how the assessment is integrated into battle rhythm/ feedback mechanism
				• Vet and staff the draft assessment plan
					Approved assessment plan
					Data collection plan
					Approved contingency plan/operation order

Collect Information and Intelligence	Execution	• Intelligence analysts • Current operations • Assessment element • Subordinate commanders • Interagency and multinational partners • Others, as required	Approved assessment plan Data collection plan Approved contingency plan/operation order	• JIPOE • Staff estimates • IR management • ISR planning and optimization	Data collected and organized, relevant to joint force actions, current and desired conditions
Analyze Information and Intelligence	Execution	• Primary staff • Special staff • Assessment element	Data collected and organized, relevant to joint force actions, current and desired conditions	• Assessment working group • Staff estimates • Vet and validate recommendations	Draft assessment products Vetted and validated recommendations
Communicate Feedback and Recommend-ations	Execution	• Commander • Subordinate commanders (periodically) • Primary staff • Special staff • Assessment element	Draft assessment products Vetted and validated recommendations	• Provide timely recommendations to appropriate decision makers	Approved assessment products, decisions, and recommendations to higher headquarters
Adapt Plans or Operations/ Campaigns	Execution Planning	• Commander • Planners • Primary staff • Special staff • Assessment element	Approved assessment products, decisions, and recommendations to higher headquarters	• Develop branches and sequels • Modify operational approach/plan • Modify objectives, effects, tasks • Modify assessment approach/plan)	Revised plans or fragmentary orders Updated assessment plan Updated data collection plan

Repeat Steps 3-6 until operation terminated/replaced/transitioned.
(Adjust using steps 1 and 2 as required during execution.)

Legend

CIPG	commander's initial planning guidance	JIPOE	joint intelligence preparation of the operational environment
CONOPS	concept of operations		
IR	intelligence requirement	JPP	joint planning process
ISR	intelligence, surveillance, and reconnaissance	OE	operational environment

Figure VI-3. Operation Assessment Steps

significantly revised strategic direction, assessment roles and responsibilities should be identified and understood. Identifying an experienced assessment development lead can help planners throughout mission analysis, CCIR development, success criteria, COA

Step 1 – Develop Operation Assessment Approach

Inputs	Staff Activity	Outputs
Strategic Guidance CIPG • Description of OE • Problem to be solved • Operational approach • Commander's intent (purpose, end state, risk)	• Conduct JIPOE • Develop operational approach • Support development and refinement of end states, objectives, effects, and tasks • Conduct joint planning (JPP and operational design) • Determine and develop how to assess tasks, effects, objectives, and end state progress for each course of action: • Identify indicators	Assessment approach which includes: framework for assessment and construct, and identification of reporting SME for MOEs and MOPs. Specific outcomes (end state, objectives, effects) Commander's estimate/ CONOPS (from JPP)

Legend

CIPG	commander's initial planning guidance	MOE	measure of effectiveness
CONOPS	concept of operations	MOP	measure of performance
JIPOE	joint intelligence preparation of the operational environment	OE	operational environment
		SME	subject matter expert
JPP	joint planning process		

Figure VI-4. Step 1—Develop Operation Assessment Approach

selection, and eventual plan development. Forming an assessment team of the right subject matter experts across the key staff, as well as inclusion of interagency and multinational partners, encourages transparency, unity of effort, and avoids duplicative efforts. Additionally, focusing on future assessment requirements throughout planning and execution can ensure the anticipated and executed tasks and objectives are assessable and help establish a logical hierarchy from measured task completion status and the creation of effects toward achievement of objectives in support of attaining the end states. Before any assessment development occurs, the purpose of the assessment and key decisions to inform throughout planning and execution should be understood.

(a) This step is focused on the linkage with the planners to ensure the assessment approach is developed as the plan and operational design is developed, and appropriately nested with the operational design. As end states, objectives, desired effects, decision points, and tasks are identified, the assessments team should determine the right measures and indicators to inform the collection effort. The ever-changing OE requires continuous monitoring and adjustment of the plan. The operation assessment complements the APEX process in answering three primary questions: Where are we? How did we get here? And what's next? The complexity of an operation often makes it difficult to determine the criteria of success. Constructed properly, the assessment plan enables appropriate monitoring and collection of necessary information and intelligence to inform critical decisions throughout planning and execution. The assessment does not replace but rather complements the commander's intuition enabled through battlefield circulation and discussions with subordinate leaders.

(b) The goal of operation assessment is to enhance the effectiveness of planning and execution by identifying and measuring observable key indicators towards progress or regression, and providing recommendations to senior decision makers to correct deficiencies and exploit success. A major challenge for the assessment team is to understand "how much is enough," in order not to overstress finite collection assets.

(c) The assessment approach should identify the information and intelligence needed to assess progress and inform decision points throughout planning and execution. This information and intelligence should be included in the CCIRs and should provide the basis for identifying changes in conditions within the OE related to specific objectives, or end states. Because success and termination criteria require identifying specific conditions within the OE, assessment considerations should be part of their development. Integration, refinement, and adaptation of assessment requirements throughout planning and execution help ensure decision points, objectives, and tasks and enhance the effective allocation and employment of joint capabilities.

(d) Once data requirements have been identified, collection requirements must be established. Essentially, the organization must consider the rules and content of the data collection plan (DCP)—the "who, where, when, how, and who (again)" (see Figure VI-5). Collection requirements should be developed with functional subject matter experts, vetted throughout the staff, and included in the DCP.

(3) These broad actions typically fall under one of three categories: organize and collect, analyze, and communicate.

(a) **Organize and Collect**

1. **Organizing for assessment involves identifying the information needed to assess effectiveness throughout planning and execution.** In the case of assessment, the required information should promote understanding of the OE in order to assess the difference between present and desired OE conditions toward achieving the

---

Information Collection Considerations

Commands should incorporate the following considerations when developing collection plans:

- Identify the information requirements that will indicate progress or regression regarding the objective or end state.
- Identify the information requirements that will inform critical decisions throughout planning and execution.
- Identify the organization or individual responsible for collecting the data.
- Determine how often the data should be updated based on the scope of the campaign or operation and changes to the operational environment.

Figure VI-5. Information Collection Considerations

---

objective or attaining the end state, and to assess the performance and effects of completed tasks and missions.

2. **Identifying the required information must be an integral part of joint planning.** Assessment considerations should be informed and shaped by operational design and reflect the operational approach. Because success and termination criteria require identifying specific conditions within the OE, assessment considerations should be part of their development. Integration, refinement, and adaptation of the required information throughout planning and execution help make certain that recommendations regarding the effectiveness of anticipated and completed tasks and missions are synchronized to support decision points.

3. **Once information requirements have been identified, it should be determined whether the required information exists or needs to be collected and collection requirements established.** Information requirements should be vetted throughout the staff and included in the initial and subsequent CCIRs. This establishes an OE baseline against which the effectiveness of anticipated and completed operations can be compared. While often considered part of execution, collection efforts for assessment should begin during planning. This process continues until planning or execution is terminated.

(b) **Analyze.** Analysis identifies operationally significant trends and changes to the OE; their impact on planning an execution (including risks and opportunities); and develops recommendations to refine, adapt, or terminate planning and execution. Information considerations are included in the development of the assessment approach and included in the assessment plan. Command internal analysis begins with plan initiation and continues throughout planning and execution until termination. External analysis support typically follows plan or order approval, but informal collaboration with supporting CCMDs, Services, and DOD agencies should begin at plan initiation.

(c) **Communicate.** The communicate action provides appropriate assessment products to all stakeholders and interested audiences. Internal communication includes the commander (for decisions regarding the overall operation or areas where the commander has expressed interest), staff elements and subordinate commands requiring information related to the analyses (for additional analysis or internal functional decisions), and external audiences (whose products may require commander approval for release). Communication considerations should be addressed during development of the assessment approach and included in the assessment plan. However, the "communicate" action typically follows plan or order approval (e.g., CONPLAN, OPLAN) and analysis.

(4) During planning for an operation or campaign, a baseline understanding of the OE assists the commander and staff in setting objectives, if useful, for desired rates of change within the OE and thresholds for success and failure. This also focuses the assessment process on answering specific questions relating to the desired objectives of the plan.

(a) Identifying the desired objective or end state and the associated conditions is critical to determining progress in any operation or campaign. Poorly defined objectives or end states typically result in ineffective planning and execution. Poorly defined objectives and end states increase the risks of wasting time, resources, and opportunities to successfully accomplish the mission. To avert this, the staff should identify clear objectives and tasks having performance and effects criteria than can be observed and measured, refined, and adapted throughout planning and execution. In turn, analysis of anticipated and completed tasks should generate assessment recommendations to communicate.

(b) Throughout planning and execution the command defines the desired observable changes in the OE necessary to accomplish the objective or end state, and may develop a DSM. The DSM is a written record of a wargamed COA that describes decision points and associated actions at those decision points. Among other information, the DSM lists decision points and the criteria to be evaluated at decision points. It also lists the units responsible for observing and reporting information affecting the criteria for decisions. This information should be incorporated into the assessment plan and reflect the required information considerations in Figure VI-6.

(5) Nonmilitary aspects of the OE may be critically important in some operations. Information derived from multiple external sources should contribute to tailored JIPOE products that address the relevant nonmilitary actors and relationships within the OE.

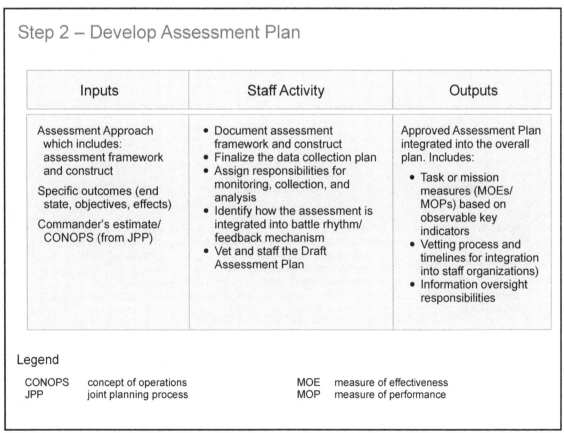

Figure VI-6. Step 2—Develop Assessment Plan

Analysis addressing all relevant actors within the OE improves and informs the understanding of how conditions may be changed within the OE.

For a detailed discussion, see Section C, "Linking Effects, Objective, and End States to Tasks Through Indicators."

b. **Step 2—Develop Operation Assessment Plan**

(1) Developing, refining, and adapting the assessment plan is concurrent and complementary throughout joint planning and execution. This step overlaps with the previous step during ID of the objectives and effects. Developing the assessment plan is a whole of staff effort and should include other key stakeholders to better shape the assessment effort. The assessment plan should identify staff or subordinate organizations to monitor, collect, analyze information, and develop recommendations and assessment products as required. Requirements for staff coordination and presentation to the commander should also be included in the plan and integrated into the command's battle rhythm to support the commander's decision cycle (Figure VI-6).

(2) The assessment plan should link objectives or end states to task or mission completion performance and effects based on observable key indicators. It should include required information oversight responsibilities to gather, update, process and exploit, analyze and integrate, disseminate, classify, and archive the required information.

c. **Step 3—Collect Information and Intelligence**

(1) Commands should collect relevant information throughout planning and execution (see Figure VI-7).

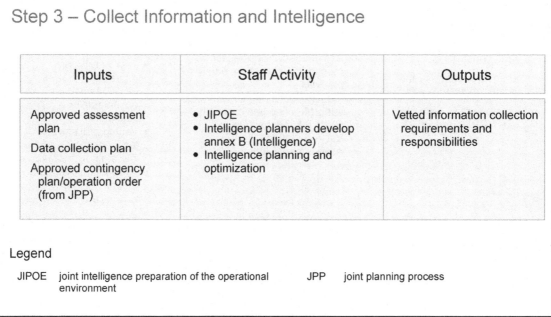

Figure VI-7. Step 3—Collect Information and Intelligence

(2) Throughout planning and execution the joint force refines and adapts information collection requirements to gather information about the OE and the joint force's anticipated and completed actions as part of normal C2 activities. Typically, staffs and subordinate commands provide information about planning and execution on a regular cycle through specified battle rhythm events. Intelligence staffs continually provide intelligence about the OE and operational impact to support the collective staff assessment effort. In accordance with the assessment plan, assessment considerations may help the staff determine the presence of decision point triggers and other mission impacts.

d. **Step 4—Analyze Information and Intelligence**

(1) Accurate, unbiased analysis seeks to identify operationally significant trends and changes to the OE and their impact on the operation or campaign. To increase credibility and transparency, **analysis should be conducted with and vetted through functional experts within the staff.** Some assessment elements may lack the expertise required to judge the impact to a particular functional area.

(2) Based on analysis, the staff can estimate the effects of force employment and resource allocation, determine whether objectives are being achieved, or determine if a decision point has been reached. Using these determinations, the staff also may identify additional risks and challenges to mission accomplishment or identify opportunities to accelerate mission accomplishment (see Figure VI-8).

(a) To identify trends and changes, it is necessary to collect and analyze observable key indicators of those differences in conditions in the OE that are the result of completed tasks and missions rather than simply OE noise or normal variation. Analysis

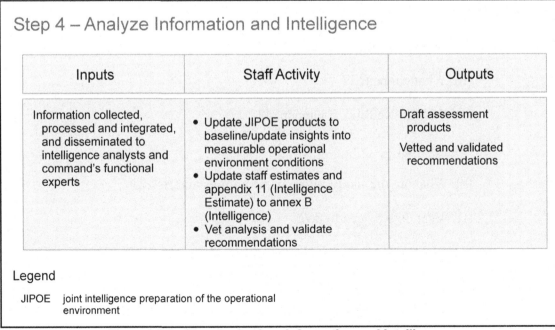

Figure VI-8. Step 4—Analyze Information and Intelligence

seeks to identify positive or negative movement toward creating desired effects, achieving objectives, or attaining end states.

(b) While individual staff elements may be responsible for analysis within their functional area, vetting and validation across the staff should enable coherent, holistic assessment products that reflect and encompass numerous discreet analyses.

(c) **Cautionary Notes**

1. Military operations are inherently human endeavors. In contrast, models are abstract representations of the OE, limited by the perspectives of their developers and what their users are attempting to evaluate. Consequently, models may not include all the critical variables and relationships in the OE. The presence of numbers or mathematical formulae in an assessment does not imply deterministic certainty, rigor, or quality. However, despite the inherent presence of uncertainty in modeling outputs, with the appropriate supporting context, assessment models can assist in analyzing and understanding complex, ill-structured OEs. Models can assist planners and assessors in producing assessments by providing objective, rational, structured approaches towards complex systems and issues.

2. Military units often find stability activities the most challenging to assess accurately. Staff elements should use caution when seeking to quantify data related to social phenomena. They should ensure military and nonmilitary subject matter experts validate data quality and its appropriateness to the phenomena and answers being sought. This type of data normally requires a sound statistical approach and expert interpretation to be meaningful in analysis.

(3) Using professional military judgment, the assessment describes progress or regress toward attaining the end state, achieving the objectives, decisive conditions, and creating effects by answering the assessment-essential questions:

(a) Where are we?

(b) What happened?

(c) Why do we think it happened?

(d) So what?

(e) What are the likely future opportunities and risks?

(f) What do we need to do?

(4) The conclusions generated by the staff analyses regarding achievement of the objective or attainment of the desired end state, force employment, resource allocation, validity of planning assumptions, and decision points should lead the staff to develop recommendations for consideration. Recommendations should highlight ways to improve

the effectiveness of operations and plans by informing all decisions, including the following:

(a) Update, change, add, or remove critical assumptions.

(b) Transition between phases (as appropriate).

(c) Execute branches and sequels.

(d) Realign resources.

(e) Adjust operations.

(f) Adjust orders, objectives, and end states.

(g) Adjust priorities.

(h) Change priorities of effort.

(i) Change support commands.

(j) Adjust command relationships.

(k) Adjust decision points.

(l) Refine or adapt the assessment plan.

(5) Before recommendations are presented to the commander for action, they must be vetted and validated through the staff. The assessment plan should detail the staff processes required to make certain assessment products are valid and any associated recommendations are achievable and improve the effectiveness of operations. A notional example of battle rhythm activities used to vet and validate assessment products is found in Figure VI-9. Many recommendations will involve decision makers below the commander. Those recommendations, once vetted and validated, should be implemented at the appropriate level. Remaining recommendations, including contentious issues, should be presented to the commander for approval and implementation guidance.

e. **Step 5—Communicate Feedback and Recommendations**

(1) The staff may be required to develop assessment products (which may include summary reports and briefings) containing recommendations for the commander based upon the guidelines set forth in the assessment plan. The commander's guidance is the most critical step in developing assessment products. Regardless of quality and effort, the assessment process is useless if the communication of its results is deficient or inconsistent with the commander's personal style of digesting information and making decisions.

(2) Assessment products are not the assessment itself. Neither are they the data collected for analysis. Assessment products serve the functions of informing the commander about current and anticipated conditions within the OE, evaluate the ability of

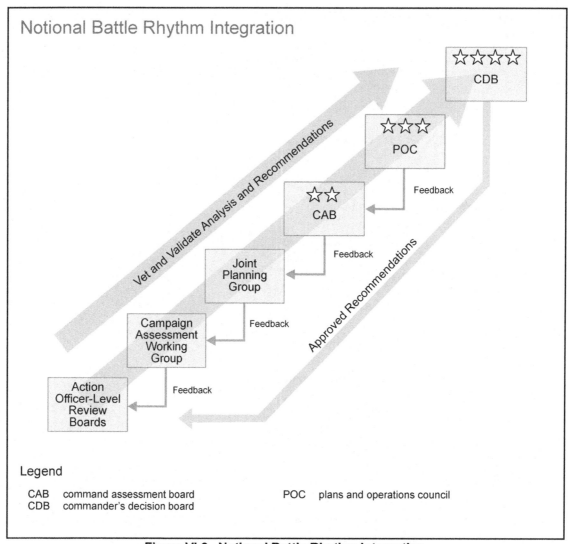

Figure VI-9. Notional Battle Rhythm Integration

the joint force to impact the OE, evaluate progress toward intermediate objectives and end states, provide accountability to higher authority, and communicate progress to multinational and interagency partners.

(3) Staffs should strive to align their efforts when communicating assessment results and recommendations (see figure VI-10). Inclusion of various staff products may gain efficiencies by possibly eliminating duplicative briefings and decision boards. It also serves to convey proper context and assure staff-wide dialogue with the commander.

f. **Step 6—Adapt Plans or Operations/Campaigns**

(1) Once feedback and recommendations have been provided, commanders typically direct changes or provide additional guidance that dictate updates or modifications to operation or campaign plan. The commander's guidance may also induce modifications to the assessment plan (see Figure VI-11). Even without significant changes

Step 5 – Communicate Feedback and Recommendations

Inputs	Staff Activity	Outputs
Draft assessment products Vetted and validated recommendations	• Provide timely recommendations to appropriate decision makers through decision boards	Approved assessment products, decisions, and recommendations to higher headquarters

Figure VI-10. Step 5—Communicate Feedback and Recommendations

to the plan or order, changes to the assessment plan may be necessary to reflect changes in the OE or adjustments to the information or intelligence requirements.

(2) As the operation or campaign transitions between phases (if applied), the assessment plan will likely require updates to adjust to changes in objectives, effects, and tasks associated with the new phase. While some of these changes can be anticipated during the original assessment plan development, revisions may be necessary to reflect actual conditions in the OE or changes to the plan or order.

(3) There should be organizational procedures associated with capturing the commander's decisions and guidance to ensure necessary actions are taken. The on-going assessment process should account for these decisions and the actions taken.

Step 6 – Adapt Plans for Operations, Campaigns, and Assessment

Inputs	Staff Activity	Outputs
Approved recommendations from the assessment process	• Develop branches and sequels • Modify operational approach • Modify objectives, effects, tasks • Modify MOPs and MOEs, information and intelligence requirements, and indicators (as necessary)	Revised plans or fragmentary orders

Legend

MOE   measure of effectiveness   MOP   measure of performance

Figure VI-11. Step 6—Adapt Plans for Operations, Campaigns, and Assessment

8. Cyclical Nature of Assessment

Until the end state has been attained or the objectives have been achieved, or the operation is terminated or transitioned, operation assessment remains an on-going process.

a. Adjustments to the plan or order based on commander's updated guidance or changes within the OE will require similar updates or changes to the assessment plan and perhaps its DCP. Updates to the plan or order should be formalized as FRAGORDs for the widest possible dissemination. Each completed analysis will identify new baseline conditions for the OE and the new basis for analyses of progress.

b. If the operation is incorporated into the command's campaign plan, appropriate intelligence and information requirements should be incorporated into the CCIRs and the assessment plan (and perhaps its DCP) for the campaign plan.

c. Once the plan or operation is terminated or when refined or adapted, commands should document their assessment approach and results as part of the lessons learned process.

SECTION C. LINKING EFFECTS, OBJECTIVES, AND END STATES TO TASKS THROUGH INDICATORS

9. Introduction

a. An operation's desired effects, objectives, and end states should help focus the staff's assessment efforts by identifying and analyzing a subset of the overall changes within the overall OE. As the staff develops the desired effects, objectives, and end states during planning, they should concurrently identify the specific pieces of information needed to infer changes in the OE supporting them. These pieces of information are commonly referred to as indicators.

KEY TERM

Indicator: In the context of operation assessment, a specific piece of information that infers the condition, state, or existence of something, and provides a reliable means to ascertain performance or effectiveness.

---

b. Indicators share common characteristics with carefully selected MOPs and MOEs and link tasks to effects, objectives, and end states (see Figure VI-12). Commanders and staffs should develop an approach that best fits their organization, operation, and requirements.

10. Guidelines for Indicator Development

a. Indicators should be **relevant, observable or collectable, responsive,** and **resourced.**

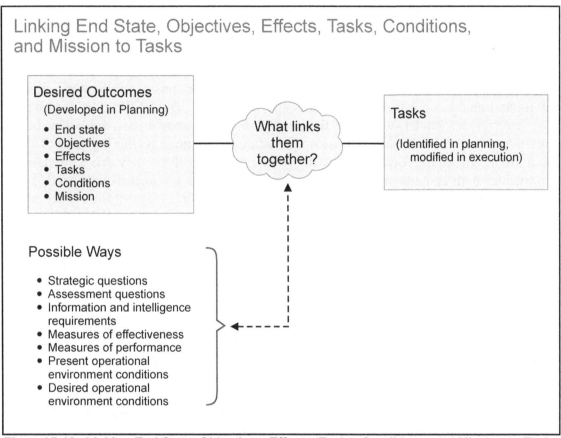

Figure VI-12. Linking End State, Objectives, Effects, Tasks, Conditions, and Mission to Tasks

(1) **Relevant.** Indicators should be relevant to a desired effect, objective, or end state within the plan or order. A valid indicator bears a direct relationship to the desired effect, objective, or end state and accurately signifies the anticipated or actual status of something about the effect, objective, or end state that must be known. This criterion helps avoid collecting and analyzing information that is of no value to a specific operation. It also helps ensure efficiency by eliminating redundant efforts.

(2) **Observable and Collectable.** Indicators must be observable (and therefore collectable) such that changes can be detected and measured or evaluated. The staff should make note of indicators that are relevant but not collectable and report them to the commander. Collection shortfalls can often put the analysis quality at risk. The commander must decide whether to accept this risk, realign resources to collect required information, or modify the plan or order.

(3) **Responsive.** Indicators should signify changes in the OE timely enough to enable effective response by the staff and timely decisions by the commander. Assessors must consider an indicator's responsiveness to stimulus in the OE. If it reacts too slowly, opportunities for response are likely to be missed; if too quickly, it exposes the staff and commander to false alarms. The JFC and staff should consider the time required for a task or mission to produce desired results within the OE and develop indicators that can respond accordingly. Many actions directed by the JFC require time to implement and may take even longer to produce a measurable result.

(4) **Resourced.** The collection of indicators should be adequately resourced so the command and subordinate units can obtain the required information without excessive effort or cost. Indicator information should be derived from other staff processes whenever possible. Assessors should avoid indicators that require development of an additional collection system. Staffs should ensure resource requirements for indicator collection efforts and analysis are included in plans and monitored. Data collection and analysis requirements associated with the threat and the OE should be embodied in the commander's PIRs with relevant tasks specified through Annex B (Intelligence) to a plan or an order. Given the focus of PIRs, the collection and analysis they drive provides the commander with insights on changes associated with MOEs. On the other hand, FFIRs provide insights to the commander on the ability of major force elements and other critical capabilities to execute their assigned tasks. Thus, they are associated with MOPs and should be published in Annex C (Operations) with reporting requirements and procedures specified in Annex R (Reports). Effective assessment planning can help avoid duplicating tasks and unnecessary actions, which in turn can help preserve combat power.

b. Collection plans must clearly articulate why an indicator is necessary for the accurate assessment of an action. Collection may draw on subordinate unit operations, KLEs, joint functions and functional estimates, and battle damage assessment. Staffs need to understand the fidelity of the available information, choose appropriate information, and prioritize use of scarce collection resources.

c. Some assessment indicators must compete for prioritization and collection assets. Assessors should coordinate with intelligence planners throughout planning and execution to identify collection efforts already gathering indicator information, alternative indicator information that might be available, and coordinate and synchronize assessment-related collection requirements with the command's integrated collection plan.

11. Selecting Indicators

a. The two types of indicators commonly used by the joint forces are MOPs and MOEs.

(1) MOPs are indicators used to assess friendly (i.e., multinational) actions tied to measuring task accomplishment. MOPs commonly reside in task execution matrices and confirm or deny proper task performance. MOPs help answer the question, "Are we doing things right?" or "Was the action taken?" or "Was the task completed to standard?"

(2) MOEs are indicators used to help measure a current system state, with change indicated by comparing multiple observations over time to gauge the achievement of objectives and attainment of end states. MOEs help answer the question, "Are we doing the right things to create the effects or changes in the conditions of the OE that we desire?"

b. Choose distinct indicators. Using indicators that are too similar to each other can result in the repetitious evaluation of change in a particular condition. In this way, similar indicators skew analyses by overestimating, or 'double-counting,' change in one item in the OE.

c. Include indicators from different causal chains. When indicators have a cause and effect relationship with each other, either directly or indirectly, it decreases their value in measuring a particular condition. Measuring progress toward a desired condition by multiple means adds rigor to the analyses.

d. Use the same indicator for more than one end state, objective, effect, task, condition, or mission when appropriate. This sort of duplication in organizing OE information does not introduce significant bias unless carried to an extreme.

e. Avoid or minimize additional reporting requirements for subordinate units. In many cases, commanders may use information generated by other staff elements as indicators in the assessment plan. Units collect many assessment indicators as part of routine operational and intelligence reporting. With careful consideration, commanders and staffs can often find viable alternative indicators without creating new reporting requirements. Excessive reporting requirements can render an otherwise valid assessment plan untenable.

f. Maximize clarity. An indicator describes the sought-after information, including specifics on time, information, geography, or unit, as necessary. Any staff member should be able to read the indicator and precisely understand the information it describes.

12. Understanding Information Categories and Data Types

a. **Information Categories.** The specific type of information that is expressed in indicators can typically be categorized as quantitative or qualitative, and subjective or objective.

(1) Since these four terms are susceptible to misinterpretation regarding assessments the following provides a guide to their meanings:

(a) **Quantitative.** Numerical information relating to the quantity or amount of something.

(b) **Qualitative.** Information reflecting an observation of, relating to, or involving quality or kind, that is typically expressed as a word, a sentence, a description, or a code that represents a category.

(c) **Subjective.** Information that is based on an individual interpretation of an observed item or condition.

(d) **Objective.** Information based on facts and the precise measurement of conditions or concepts that actually exist without distortion by personal feelings, prejudices, or interpretations.

(2) To ensure value and credibility, assessors must understand and apply categorization considerations in their assessments and recommendations. Indicator information is usually a combination of the four information categories: quantitative-objective, quantitative-subjective, qualitative-objective, and qualitative-subjective (as

shown in Figure VI-13). As a standard of analytical rigor, information category must be considered when formulating analyses, reports, and recommendations.

b. **Information Types.** Assessment information is used to calculate, analyze, and recommend. Whenever possible, information should be empirical—originating in or based on observation or experience. Generally, there are four information types. Knowing the type is essential to understand the type of analysis that can be performed, and whether the information can be interpreted to draw conclusions, such as the quantity and speed of change in an OE condition over time. In increasing level of complexity and information content they are:

(1) **Nominal.** Nominal information can be organized or sorted into categories, with no difference in degree or amount between category and any ordering by category is arbitrary. For example, friendly forces are categorized by sending nation (e.g., from Albania, Belgium, Bulgaria).

(2) **Ordinal.** Ordinal information has an order, but does not indicate the magnitude of discrete intervals within the information. A Likert Scale is a common application of ordinal information, where "strongly agree" represents more agreement than "agree," but without specifying how much more. An example of ordinal data might include the rating of the capability of a unit from "able to perform independent operations" as the highest and "unable to perform operations without assistance" as the lowest rating.

(3) **Interval.** Interval information is ordinal data with the extra property of having the discrete intervals qualified, or able to be meaningfully added or subtracted. However, an interval scale has no meaningful value for zero, so ratios are meaningless. An example is temperature scales, where 0°Celsius does not mean that there is no temperature. To illustrate, the average daily temperature in Kabul in June may be 25°Celsius, and in December, 5°Celsius; so, while a difference of 20°Celsius between these months is meaningful, it cannot be stated that June is 5 times as hot as December.

Information Category Example		
	Quantitative	Qualitative
Objective	The number of no-fly zone violations that have occurred in the last week	The mandate to enforce a no-fly zone is approved
Subjective	The air component's assessment of the effectiveness of the no-fly zone, on a scale of 1 to 10	Enemy freedom of action is limited by the no-fly zone

Figure VI-13. Information Category Example

(4) **Ratio.** Ratio information has meaning in both intervals and ratios between measurements. Ratio information has a natural zero, indicating the absence of whatever is being measured. For example, the number of personnel in the armed forces of NATO nations (1999 figures, in thousands) is US, 1372; Turkey, 639; Germany, 322; and so on. It is valid to say both that Turkey has 317,000 more military personnel than Germany, and that the US has more than twice as many military personnel as Turkey.

13. Linking Effects, Objectives, and End States to Tasks Through Indicators

Ensuring effects, objectives, and end states are linked to tasks through carefully selected MOPs and MOEs is essential to the analytical rigor of an assessment framework. Establishing strong, cogent links between tasks and effects, objectives, and end states through MOPs and MOEs facilitates the transparency and clarity of the assessment approach. Additionally, links between tasks and effects, objectives, and end states assist in mapping the plan's strategy to actual activities and conditions in the OE and subsequently to desired effects, objectives, and end states. The following notional example presents an approach that links tasks to effects, objectives, and end states through MOPs and MOEs. It does not reflect any current real world assessment plan or approach.

a. **Approach 1**—Using Assessment Questions and Information and Intelligence Requirements. This approach uses the model shown in Figure VI-14 to guide the development of assessment questions and information and intelligence requirements in order to identify indicators.

(1) **Statements** about effects, objectives, or end states can refer to anything that specifies the change(s) in the OE being sought. Within Figure VI-14, the refinement of a statement into "smaller statements" refers to any statement or question that increases the specificity of the original statement. For example, for a military end state, we may have several objectives; for an objective, we may have several effects; or, for a strategic objective, we may have several termination criteria. During this portion of the process, assessors help develop specific desired effects, objectives, or end states. These nested operational design elements may have one or more associated assessment questions.

(2) **Assessment questions** are those that, when answered, provide the commander and staff with direct answers to critical information pertaining to the OE and progress toward desired effects, objectives, or end states. Assessment questions take the general form of "How well are we creating our desired effects?" and related questions such as, "How can we achieve our objectives more effectively—more quickly, qualitatively better, at less cost, or at less risk?" Answers to the CCIRs should ground the assessment process in the desired effects. They should be **answerable** with the information or data available to the command; **relevant** to the desired effects, objectives, or end states and commander priorities; and **useful** to evaluate whether the mission is being performed, desired effects are being created, objectives are being achieved, and end states are being attained.

(3) **Information and intelligence requirements** should be related to the desired effects, objectives, or end states of the plan and should be developed from the assessment

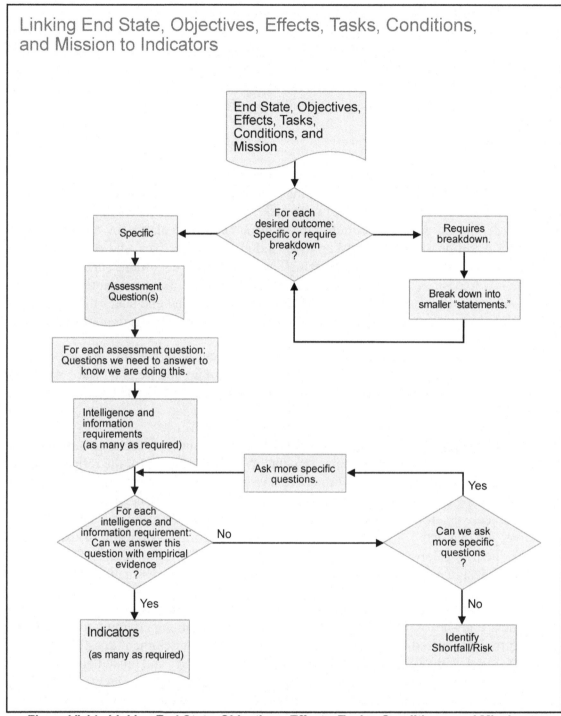

Figure VI-14. Linking End State, Objectives, Effects, Tasks, Conditions, and Mission to Indicators

questions. Information and intelligence requirements provide a foundation for the development of indicators and record the logical connection between indicators and assessment questions and the effects, objectives, or end states they support. Within the context of assessments, intelligence requirements are typically used to understand conditions within OE while information requirements are used to determine whether the joint force properly executed planned actions. By using both intelligence and information, assessment can provide more comprehensive analyses of the current OE and the joint force's impact on it (see Figure VI-15). When developing information and intelligence requirements, here are some of the questions the staff may ask to determine the value of proposed requirements:

(a) **Usage.** What aspect of the desired effects, objectives, or end states does this information or intelligence requirement inform?

(b) **Source.** How will the required information or intelligence be collected? Who is collecting it? What is our confidence level in the reporting?

(c) **Measurability.** Is the information or intelligence requirement measurable? If the information or intelligence requirement is unavailable, are there other information or intelligence requirements that can serve as proxies?

(d) **Impact.** What is the impact of knowing the required information or intelligence? What is the impact of not knowing it? What is the risk if it is false?

(e) **Timeliness.** When is the required information or intelligence no longer valuable?

Comparison and Use of Information and Intelligence

	Information	Intelligence
Perspective	Internal focus	External focus
Sources	Staff section and subordinate command reports, host-nation reports, nongovernmental organization information.	All-source intelligence, intelligence agency reports, host-nation reports.
Use in Plans	Friendly force information requirements assumptions linking force posture to operations.	Priority intelligence requirements assumptions linking operations to effects.
Use in Assessments	Identifies if planned actions are executed properly.	Identifies if desired outcomes are achieved.
Result of Assessment	Resource efficiency of the plan.	Resource effectiveness of the plan.
Example of Information or Intelligence Requirement	• Allocation of coalition trainers to train host-nation security forces within a specific region. • Readiness assessment of host-nation security forces.	Security assessment within a particular region.

Figure VI-15. Comparison and Use of Information and Intelligence

(f) **Cost.** What is the cost of data collection to answer the information and intelligence requirements (e.g., the risk to forces, resources, and or mission)?

(4) Indicators should answer information and intelligence requirements. Indicator characteristics are discussed in paragraph 10, "Guidelines for Indicator Development."

b. **Notional Example of Approach 1**

(1) A headquarters has established the end state, "Professional and Self-Sustaining Security Institutions are created." Planners, working with assessors, review the end state and develop more specific objectives.

(2) The staff then develops the initial assessment international and intelligence requirements from the effects or objectives (see Figure VI-16). Note that initial international and intelligence requirements essentially mirror the more specific statements.

(3) Assessors continue to refine the initial international and intelligence requirements to develop more specific international and intelligence requirements, as shown in Figure VI-17.

(4) At this point the staff, assisted by the assessors, attempt to develop more specific international and intelligence requirements. For example, for the smaller assessment question, "Are there requirements that specify skills the soldiers need to have?" the staff may develop the information requirements as shown in Figure VI-18.

Developing Assessment Questions from Smaller Outcome Statements (Example)

Objectives	Assessment Questions
Security forces are properly manned.	Are security forces properly manned?
Security forces are properly trained.	Are security forces properly trained?
Security forces are properly equipped.	Are security forces properly equipped?
Security forces are properly sustained.	Are security forces properly sustained?
Security forces are effective in exercise. Security forces are effective in combat.	Are security forces effective in exercise? Are security forces effective in combat?
Security forces are accountable to legitimate authority.	Are security forces accountable to legitimate authority?
Security forces have institutional infrastructure for sustaining 1-6.	Do security forces have institutional infrastructure for sustaining 1-6?

Figure VI-16. Developing Assessment Questions from Smaller Outcome Statements (Example)

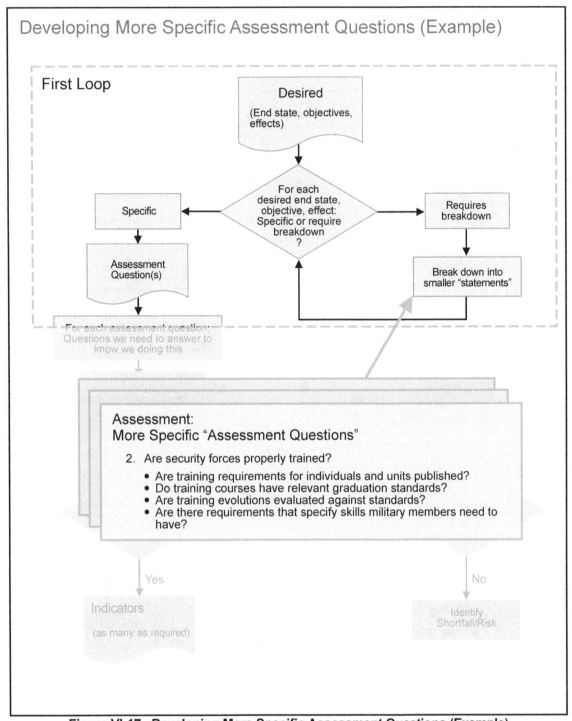

Figure VI-17. Developing More Specific Assessment Questions (Example)

(5) Once the international and intelligence requirements have been identified, the staff begins to identify appropriate indicators that answer the international and intelligence requirements. If a required indicator cannot be identified or cannot be observed, it should be identified as a shortfall and reported in the assessment plan. In addition to including it in the assessment plan, **the commander should be informed of the shortfall and its potential impact on the assessment and, more importantly, the**

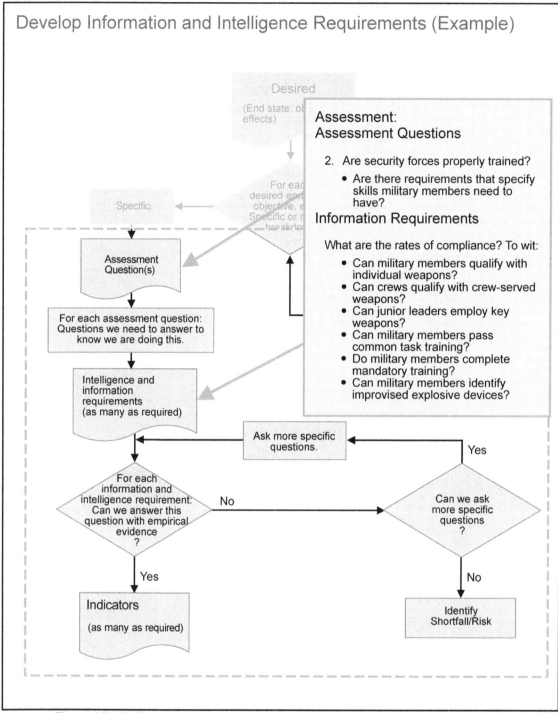

Figure VI-18. Develop Information and Intelligence Requirements (Example)

overall operation (see Figure VI-19). Once the indicators are developed, the staff should develop the DCP and include a record of the process in the assessment plan.

c. **Approach 2**—Develop indicators to assess operations. This approach facilitates the development of MOPs and MOEs (see Figure VI-20).

Develop Indicators (Example)

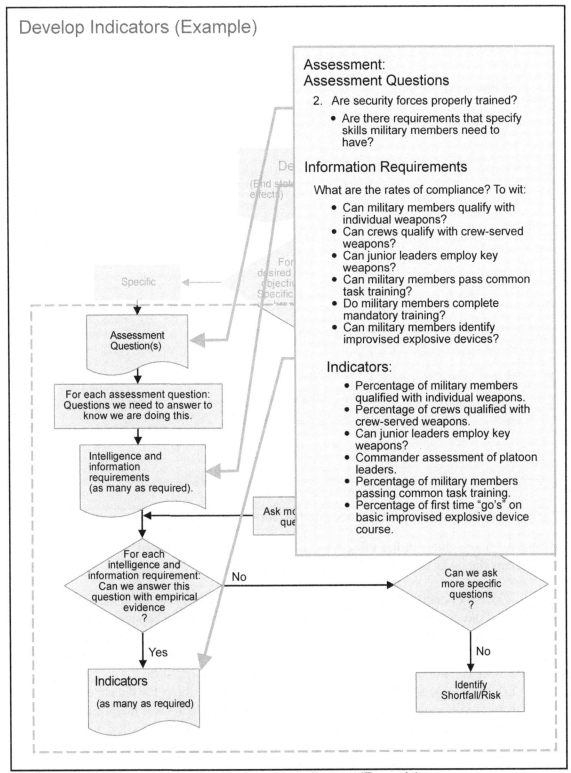

Figure VI-19. Develop Indicators (Example)

(1) During planning, the OPT, as supported by assessors, determines a hierarchy of increasingly specific or more refined statements. For example, these may be the

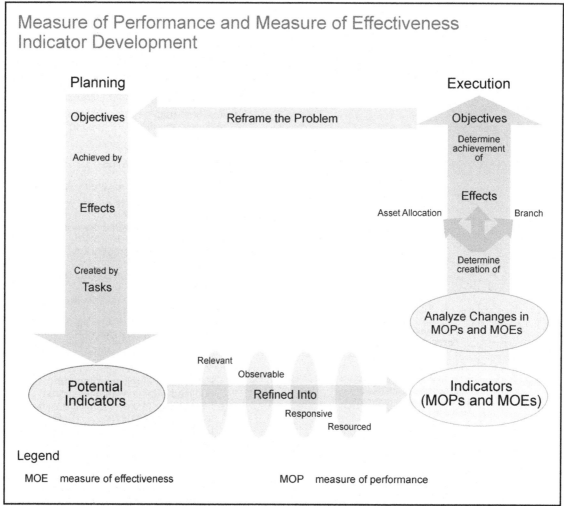

Figure VI-20. Measure of Performance and Measure of Effectiveness Indicator Development

objectives to be achieved, the effects to be created in the OE to achieve those objectives, and perhaps the tasks intended to create those effects.

(2) Functional experts, supported by assessors, then develop potential indicators for each effect. Potential indicators should answer the questions "What happened?" and "How do we know we are creating the desired effects?" The answers to these questions are indicators that may inform MOPs and MOEs. Performance-oriented indicators reflect friendly force actions and activities and inform MOP. They help answer the question, "Are we doing things right?" Effectiveness-oriented indicators reflect a current condition for the state of some part of the OE and are commonly referred to as MOEs. MOEs help answer the question, "Are we doing the right things?" The following steps present a logical process the staff can use to develop measures and indicators (either MOPs or MOEs) for each desired effect.

(a) Analyze the desired effects and tasks.

(b) Identify candidate MOPs and MOEs for subsequent refinement. Consider developing MOPs, (and MOP indicators, if used) that reflect progress in achieving key tasks as the approach to performance assessment.

(c) Refine MOEs and MOPs. They should be relevant to the desired effect (MOEs) or associated task (MOPs), observable, responsive, and resourced.

(d) Identify collection requirements for MOPs and MOEs. Requirements should be prioritized for inclusion in the command's collection plans. Since MOPs reflect friendly force actions and activities, most will be available through routine reports and should not require separate collection efforts for assessment. Some collection requirements for MOEs may also be available as part of the command's JIPOE efforts. However, some MOEs will require new collection efforts to gather the appropriate information and must compete for resources with other command collection requirements. Those indicators informing MOPs and MOEs that cannot be collected must be identified and included as part of the DCP along with the risk associated with loss of that information.

(e) Incorporate indicators into the DCP and assessment plan.

(f) Monitor and modify indicators as necessary during execution.

d. **Notional Example of Approach 2.** As part of a coalition task force, the coalition force maritime component commander has established the desired end state, "Country Green effectively controls its internationally recognized maritime territory consisting of its territorial seas and economic exclusion zone." During operational design and mission analysis, the OPT—with cross-functional representation from plans, operations, intelligence, assessment, and others—reviews the desired objectives, identifies desired conditions, and develops specific objectives as seen in Figure VI-21.

(1) The OPT continues its planning activities and begins to develop potential COAs. Each COA further refines each objective into effects and tasks (see Figure VI-22).

Develop Conditions and Objectives

End State: "Country Green effectively controls its internationally recognized maritime territory, consisting of its territorial seas and economic exclusion zone."

Conditions	Objectives
Rule of Law is enforced in Green maritime territory.	Coalition enforces Rule of Law in Green territorial seas and economic exclusion zone.
	Green enforces Rule of Law in Green territorial seas and economic exclusion zone.
Red aggression ceases in Green maritime territory.	Coalition defeats current Red aggression in Green territorial seas and economic exclusion zone.
	Green deters future Red aggression in Green territorial seas and economic exclusion zone.

Figure VI-21. Develop Conditions and Objectives

Develop Effects

End State: "Country Green effectively controls its internationally recognized maritime territory, consisting of its territorial seas and economic exclusion zone."

Conditions	Objectives	Effects
Red aggression ceases in Green maritime territory.	Coalition defeats current Red aggression in Green territorial seas and economic exclusion zone.	Current Red naval operations cease in Green territorial seas. Current Red air operations cease in Green territorial seas.
		Hostile Red naval operations cease in Green economic exclusion zone. Hostile Red air operations cease in Green economic exclusion zone.

Figure VI-22. Develop Effects

Functional staff elements, again supported by assessors, develop staff estimates for the COAs and evaluate the ability for the command to assess effects and tasks associated with each COA.

(2) In conjunction with the OPT, these functional staff elements begin to develop indicators (or MOE) for each effect using sample questions such as: "How do we know we're creating this effect?" "How can we recognize success?" and "What indicators can we use to gauge change?" These initial measures are included as part of the COA selection process and incorporated in the planning process. Figure VI-23 highlights an example of brainstorming the OPT may conduct to develop potential indicators for the effect, "Red naval operations cease in Green territorial seas."

(3) The OPT should next evaluate each potential indicator to ensure it is relevant and collectible. For example, in evaluating the potential indicator, "Red forces are moving to reposition outside Green territorial seas," the OPT should ensure it:

(a) Is relevant to the desired effect. The measure is useful to identify whether Red naval forces are leaving Green territorial seas. If all forces leave, Red naval operations in Green territorial seas will have ended.

(b) Is observable and collectible. In this case, routine intelligence monitoring and reporting normally contains this information. Analyzing reports over a period of time can provide a trend in the activity.

(c) Is understandable. The potential indicators should lead to one or more refined indicators that be easily understood by anyone reading them.

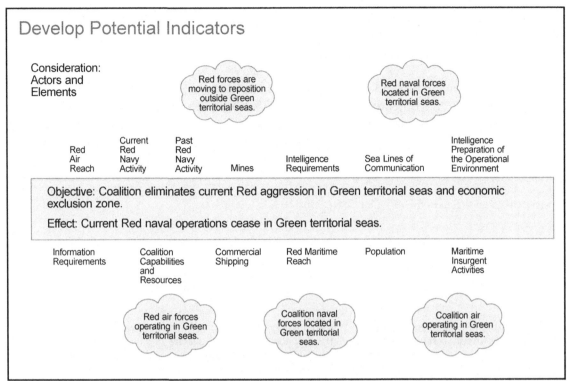

Figure VI-23. Develop Potential Indicators

(4) Potential indicators should be refined into one or more refined indicators. As discussed earlier, refined indicators should be relevant to the desired mission, condition, task, effect, objective, or end state, observable, responsive, and resourced. Figure VI-24 shows an example identifying refined indicators for the potential indicator, "Red naval forces are moving to reposition outside Green territorial seas over the past 96 hours" Note

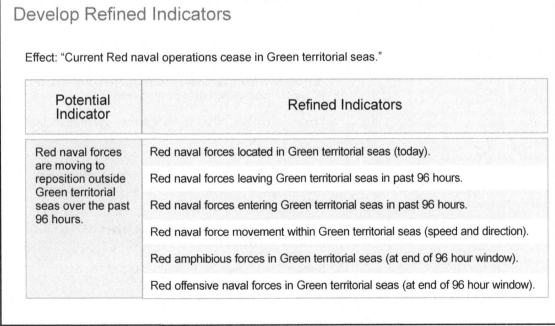

Figure VI-24. Develop Refined Indicators

the addition of a time frame in the measure. In this case, the staff, in coordination with the intelligence element, determined that observable trends would probably take about 96 hours to develop based on the current operational tempo.

(5) Once the process is completed, the staff should have a set of indicators linked to desired effects and objectives. Figure VI-25 shows an example of the relationship between a desired objective, effect, and associated indicators. Once the refined indicators have been identified and developed, the potential indicators are no longer required.

Linking Objectives and Effects to Indicators

Desired Condition: Red aggression ceases in Green maritime territory.

Objective	Effect	Indicators
Coalition defeats current Red aggression in Green territorial seas and economic exclusion zone.	Current Red naval operations cease Green territorial seas.	Red naval forces located in Green territorial seas at end of 96 hour window (measure of effectiveness). Red naval forces leaving Green territorial seas in past 96 hours (measure of effectiveness). Red naval forces entering Green territorial seas in past 96 hours (measure of effectiveness). Red naval force location(s) within Green territorial seas (measure of effectiveness). Red amphibious forces located in Green territorial seas (at end of 96 hour window) (measure of effectiveness). Red offensive naval forces located in Green territorial seas (at end of 96 hour window) (measure of effectiveness). Red offensive naval actions in Green territorial seas in past 96 hours (measure of effectiveness). Number of coalition defensive actions in Green territorial seas in past 96 hours (measure of performance). Number of coalition offensive naval actions against Red naval forces in Green territorial seas in past 96 hours (measure of performance). Number of coalition offensive air actions against Red naval forces in Green territorial seas in past 96 hours (measure of performance). Number of Red naval forces destroyed in past 96 hours (measure of effectiveness). Number of Red naval forces disabled in past 96 hours (measure of effectiveness).

Figure VI-25. Linking Objectives and Effects to Indicators

Analysis of the information resulting from collection of the indicators should identify changes in the OE and determine progress of the operation.

Intentionally Blank

CHAPTER VII
TRANSITION TO EXECUTION

"A good plan, violently executed now, is better than a perfect plan next week."

George S. Patton

1. Overview

a. Plans are rarely executed as written. Regardless of how much time and effort went into the planning process, commanders and their staffs should accept that the plan, as written, will likely need changes on execution. Often, the decision to deploy the military will be in conditions significantly different from the original planning guidance or the conditions planned. Planning provides a significant head start when called to deploy the military. Assessments and reframing the problem, if required, inform the applicability of, or necessary modifications to the plan in response to changes in the OE.

(1) Effective planning enables transition. Integrated staff effort during planning ensures the plan is a team effort and the knowledge gained across the staff in the planning process is shared and retained. This staff work assists in identifying changes in the OE and guidance, speeding transition to execution.

(2) Detailed planning provides the analysis of the adversary and the OE. The knowledge and understanding gained enables a well-trained staff to quickly identify what is different between their plan and current conditions and make recommendations based on their prior work.

(3) Detailed OPLANs (levels 3 or 4) may require more significant changes due to their specificity. Forces identified in the plan may not be available, assumptions may not be validated, and policy and strategic decisions (and the decision timeline) may have changed or not support the original concept. However, the extra time spent on analysis provides a deeper understanding of the OE, adversaries, and the technical issues with projecting forces.

(4) Less detailed plans (levels 1-2) may be more readily adaptable to execution due to their generality. However, they will require significantly more analysis (e.g., forces, transportation, logistics) to provide the detail required to enable decisions at the strategic level and ensure the plan's executability and suitability for the problem at hand.

b. The decision to execute will often be presented as an examination of options in response to a developing crisis or action by a competitor state or adversary (state or non-state) rather than a specific directive to execute a specific CONPLAN or OPLAN.

(1) If an existing plan is appropriate, the commander and staff should review and update the plan. See paragraph 3, "Transition Process," for additional information.

(2) If no existing plan meets the guidance, the commander and staff conduct crisis planning (planning in reduced timeline). More often than not, the commander and staff have conducted some previous analysis of the OE which will speed the planning process.

2. Types of Transition

a. There are three possible conditions for transitioning planning to execution.

(1) **Contingency Plan Execution**

(a) Contingency plans are planned in advance to typically address an anticipated crisis. If there is an approved contingency plan that closely resembles the emergent scenario, that plan can be refined or adapted as necessary and executed. The APEX execution functions are used for all plans.

(b) Members of the planning team may not be the same as those responsible for execution. They may have rotated out or be in the planning sections of the staff rather than the operations. This is the most likely situation where the conditions used in developing the plan will have changed, due to the time lag between plan development and execution. Staff from the planning team need to provide as much background information as possible to the operations team.

(c) The planning team should be a key participant, if not the lead, in updating the plan for the current (given) conditions. This enables the command to make effective use of the understanding gained by the staff during the planning process. The operations team should be the co-lead for the plan update to ensure they understand the decision processes and reasoning used in development of the operational approach and COAs. This will speed plan update, ease transition, and minimize the time required to revisit the issues that arose during the initial plan development.

(2) **Crisis Planning to Execution.** Crisis planning is conducted when an emergent situation arises. The planning team will analyze approved contingency plans with like scenarios to determine if an existing plan applies. If a contingency plan is appropriate to the situation, it may be executed through an OPORD or FRAGORD. In a crisis, planning usually transitions rapidly to execution, so there is limited deviation between the plan and initial execution. Planners from the command J-5 can assist in the planning process through their planning expertise and knowledge gained of the OE during similar planning efforts.

(3) **Campaign Plan Execution.** Activities within campaign plans are in constant execution.

b. Planning is conducted based upon assumed forces and resources. Upon a decision to execute, these assumptions are replaced by the facts of actual available forces and resources. Disparities between planning assumptions and the actual OE conditions at execution will drive refinement or adaption of the plan or order. Resource informed planning during plan development allows planners to make more realistic force and resource planning assumptions. Enabled by the common formats and collaborative systems, tools and processes of APEX, resource informed planning is intended to facilitate the transition of a plan or order by reducing the scope of required plan adjustments or refinements upon execution.

c. During execution, the commander will likely have reason to consider updating the operational approach. It could be triggered by significant changes to understanding of the OE and/or problem, validation or invalidation of assumptions made during planning, identifying (through continuous assessment process) that the tactical actions are not resulting in the expected effects, changes in the conditions of the OE, or the end state. The commander may determine one of three ways ahead:

(1) The current OPLAN is adequate, with either no change or minor change (such as execution of a branch)—the current operational approach remains feasible.

(2) The OPLAN's mission and objectives are sound, but the operational approach is no longer feasible or acceptable—a new operational approach is required.

(3) The mission and/or objectives are no longer valid, thus a new OPLAN is required—a new operational approach is required to support the further detailed planning.

d. Assessment could cause the JFC to shift the focus of the operation, which the JFC would initiate with a new visualization manifested through new planning guidance for an adjusted operation or campaign plan.

3. **Transition Process**

a. **Overview.** The transition from plan to execution should consider the following points. These are not meant to be exclusive and may be conducted simultaneously.

b. **Transition Requirements**

(1) Update environmental frame and intelligence analysis. Identify what has changed since plan development and how that affects the plan.

(2) Identify any changes to strategic direction or guidance. This will require dialogue with senior civilian leadership to ensure the military objectives remain synchronized with policy and strategic objectives.

(a) Confirm and update strategic objectives or end states.

(b) Confirm and update operational limitations (constraints and restraints).

(c) Validate assumptions.

(d) Review and validate assessment criteria.

1. External (strategic) assumptions, especially those dealing with policy, diplomacy, and multinational partners, should be validated as part of the plan review with

senior civilian leadership. These are usually the assumptions dictated to the command through strategic directives (GEF, JSCP, SGSs) or previous planning IPRs.

2. Internal (operational) assumptions should be validated by the staff through their update of the OE.

(e) Identify partners and allies.

(f) Identify interagency participation, actions, and responsibilities.

(3) Identify forces and resources, to include transportation. The forces assumed in planning are for planning purposes only; execution sourced forces may or may not match those assumed in planning. Execution sourcing requires a dialogue between the supported CCDR, the JS, JFPs, Services, and USTRANSCOM.

(4) Identify decision points and CCIRs to aid in decision making. Ensure consideration is taken to include lead times, to include notification and mobilization for reserve forces, transportation timelines, and JRSOI requirements. These decision points are critical for senior DOD leadership to understand when decisions should be made to enable operations and reduce risk. During this discussion, commanders and planners should identify alternative COAs and the cost and risk associated with them should decisions be delayed or deferred. Decision points should specifically address how the US might use the military in:

(a) **FDOs.** When and what FDOs should be deployed and the expected impact. The discussion should identify indicators that the FDOs are creating the desired effect.

(b) **FROs.** FROs, usually used in response to terrorism, can also be employed in response to aggression by a competitor or adversary. Like FDOs, the discussion should include indicators of their effectiveness and probability of consequences, desired and undesired.

(c) **De-Escalation.** During transition to execution, commanders should identify a means for de-escalation and steps that could be taken to enable de-escalation without endangering US forces or interests.

(d) **Escalation.** Similarly, commanders need to identify decision points at which senior leaders must make decisions to escalate in order to ensure strategic advantage, to include the expected risk associated should the adversary gain the advantage prior to US commitment.

(5) **Confirm Authorities for Execution.** Request and receive President or SecDef authority to conduct military operations. Authorities granted may be for execution of an approved plan or for limited execution of select phases of an approved plan.

(6) **Direct Execution.** The JS, on behalf of the CJCS, prepares orders for the President or SecDef to authorize the execution of a plan or order. The authorities for

execution, force allocation, and deployment are often provided separately vice in a comprehensive order. Upon approval, CCDRs and Services pass orders down the chain of command directing action ordered by higher headquarters. The following orders are some of those that may be used in the process of transitioning from planning to execution: WARNORD, PLANORD, ALERTORD, OPORD, PTDO, DEPORD, EXORD, and FRAGORD.

(a) **Contingency Plans.** The authority to execute a contingency plan may be provided incrementally. Initial execution authority may be limited to early phase activities and CCDRs should be prepared to request additional or modified execution authorities as an operation develops.

(b) **CCMD Campaign Plans.** CCMD campaign plans are in constant execution. While they are reviewed by SecDef, the authorization to execute a campaign plan does not provide complete authority for the CCDR to execute all of the individual military activities that comprise the plan. Additional CCMD coordination is required to execute the discrete military activities within a campaign plan to include: posture, force allocation, and country team coordination.

See CJCSM 3130.03, Adaptive Planning and Execution (APEX) Planning Format and Guidance, *for more information on the content and format of orders.*

c. **Impact on Other Operations.** As the plan transitions to execution, the commander and staff synchronize that operation with the rest of the CCMD's theater (or functional) campaign.

(1) The commander identifies how the additional operation will affect the campaign.

(a) Resources. Resources may be diverted from lower priority operations and activities to support the new operation. This may require modifying the campaign or adjusting objectives.

(b) Secondary effects. Adding new operations, especially combat operations, will impact the perception and effects of other operations within the AOR (and likely in other CCMD's AORs as well). Both the new operation and existing ones may need to be adjusted to reflect the symbiotic effect of simultaneous operations.

(2) The commander may require support from other CCMDs. In addition to support within the plan transitioning to execution, the CCDR may require external support to ensure continued progress toward theater or functional objectives. By using a pre-established capability (force) sharing agreement, a CCDR can gain the support needed without requiring additional JS or OSD coordination. Support from other CCMDs often require shared battle rhythm activities. Balancing the benefit of improved awareness without overburdening commanders and their staffs remains a challenge. Informal cross-CCMD, directorate-level coordination has proven beneficial and can expand when security conditions necessitate deeper coordination and synchronization. However, identifying

standardized staff organizations provides additional structure when planning and scheduling across organizational boundaries.

(3) Depending on the significance of the new operation, the CCDR may need to update the theater or functional campaign objectives. This will require a conversation with senior civilian leaders to see if the US national objectives should be adjusted given the change in the strategic landscape.

APPENDIX A
JOINT OPERATION PLAN FORMAT

SECTION A. INTRODUCTION

a. Below is a sample format that a joint force staff can use as a guide when developing a joint OPLAN. The exact format and level of detail may vary somewhat among joint commands, based on theater-specific requirements and other factors. However, joint OPLANs/CONPLANs will always contain the basic five paragraphs (such as paragraph 3, "Execution") and their primary subparagraphs (such as paragraph 3a, "Concept of Operations"). **The JPEC typically refers to a joint contingency plan that encompasses more than one major operation as a campaign plan, but JFCs prepare a plan for a campaign in joint contingency plan format.**

b. The CJCSM 3130 series volumes describe joint planning interaction among the President, SecDef, CJCS, the supported joint commander, and other JPEC members, and provides models of planning messages and estimates. CJCSM 3130.03, *Adaptive Planning and Execution (APEX) Planning Formats and Guidance,* provides the formats for joint plans.

SECTION B. NOTIONAL OPERATION PLAN FORMAT

a. Copy Number

b. Issuing Headquarters

c. Place of Issue

d. Effective Date-Time Group

e. OPERATION PLAN: (Number or Code Name)

f. USXXXXCOM OPERATIONS TO . . .

g. References: (List any maps, charts, and other relevant documents deemed essential to comprehension of the plan.)

1. **Situation**

(This section briefly describes the composite conditions, circumstances, and influences of the theater strategic situation that the plan addresses [see national intelligence estimate, any multinational sources, and strategic and commanders' estimates].)

a. **General.** (This section describes the general politico-military variables that would establish the probable preconditions for execution of the contingency plans. It should summarize the competing political goals that could lead to conflict, identify primary antagonists, state US policy objectives and the estimated objectives of other parties, and outline strategic decisions needed from other countries to achieve US policy objectives and conduct effective US military operations to achieve US military objectives. Specific items can be listed separately for clarity as depicted below.)

(1) **Assessment of the Conflict.** (Provide a summary of the national and/or multinational strategic context [JSCP, UCP].)

(2) **Policy Goals.** (This section relates the strategic guidance, end state, and termination criteria to the theater situation and requirements in its global, regional, and space dimensions, interests, and intentions.)

(a) **US/Multinational Policy Objectives.** (Identify the national security, multinational or military objectives, and strategic tasks assigned to or coordinated by the CCMD.)

(b) **End State.** (Describe the national strategic end state and relate the military end state to the national strategic end state.)

(3) **Non-US National Strategic Decisions**

(4) **Operational Limitations.** (List actions that are prohibited or required by higher or multinational authority [e.g., ROE, RUF, law of war, termination criteria].)

b. **Area of Concern**

(1) **OA.** (Describe the JFC's OA. A map may be used as an attachment to graphically depict the area.)

(2) **Area of Interest.** (Describe the area of concern to the commander, including the area of influence, areas adjacent thereto, and extending into enemy territory to the objectives of current or planned operations. This area also includes areas occupied by enemy forces who could jeopardize the accomplishment of the mission.)

c. **Deterrent Options.** (Delineate FDOs and FROs desired to include those categories specified in the current JSCP. Specific units and resources must be prioritized in terms of latest arrival date relative to C-day. Include possible diplomatic, informational, or economic deterrent options accomplished by non-DOD agencies that would support US mission accomplishment.)

See Appendix F, "Flexible Deterrent Options and Flexible Response Options," for examples of FDOs and FROs.

d. **Risk.** (Risk is the probability and severity of loss linked to hazards. List the specific hazards that the joint force may encounter during the mission. List risk mitigation measures.)

e. **Enemy Forces.** (Identify the opposing forces expected upon execution and appraise their general capabilities. Refer readers to Annex B [Intelligence] for details. However, this section should provide the information essential to a clear understanding of the magnitude of the hostile threat. Identify the adversary's strategic and operational COGs and critical vulnerabilities as depicted below.)

    (1) **Enemy COGs**

        (a) Strategic.

        (b) Operational.

    (2) **Enemy Critical Factors**

        (a) Strategic.

        (b) Operational.

    (3) **Enemy COAs** (most likely and most dangerous to friendly mission accomplishment).

        (a) General.

        (b) Enemy's End State.

        (c) Enemy's Strategic Objectives.

        (d) Enemy's Operational Objectives.

        (e) Enemy CONOPS.

    (4) **Enemy Logistics and Sustainment**

    (5) **Other Enemy Forces/Capabilities**

    (6) **Enemy Reserve Mobilization**

  f. **Friendly Forces**

    (1) **Friendly COGs.** (This section should identify friendly COGs, both strategic and operational; this provides focus to force protection efforts.)

        (a) Strategic.

        (b) Operational.

    (2) **Friendly Critical Factors**

        (a) Strategic.

        (b) Operational.

(3) **MNF**

(4) **Supporting Commands and Agencies.** (Describe the operations of unassigned forces, other than those tasked to support this contingency plan that could have a direct and significant influence on the operations in the plan. Also list the specific tasks of friendly forces, commands, or government departments and agencies that would directly support execution of the contingency plan, for example, USTRANSCOM, USSTRATCOM, Defense Intelligence Agency, and so forth.)

g. **Assumptions.** (List all reasonable assumptions for all participants contained in the JSCP or other tasking on which the contingency plan is based. State expected conditions over which the JFC has no control. Include assumptions that are directly relevant to the development of the plan and supporting plans and assumptions to the plan as a whole. Include both specified and implied assumptions that, if they do not occur as expected, would invalidate the plan or its CONOPS. Specify the mobility [air and sea lift], the degree of mobilization assumed [i.e., total, full, partial, selective, or none].)

(1) **Threat Warning/Timeline.**

(2) **Pre-Positioning and Regional Access** (including international support and assistance).

(3) **In-Place Forces.**

(4) **Strategic Assumptions** (including those pertaining to nuclear weapons employment).

h. **Legal Considerations.** (List those significant legal considerations on which the plan is based.)

(1) ROE.

(2) International law, including the law of war.

(3) US law.

(4) HN and partner nation policies.

(5) Status-of-forces agreements.

(6) Other bilateral treaties and agreements.

(7) HN agreements to include HNS agreements.

2. Mission

(State concisely the essential task[s] the JFC has to accomplish. This statement should address who, what, when, where, and why.)

3. Execution

a. **CONOPS.** (For a CCDR's contingency plan, the appropriate commander's estimate can be taken from the campaign plan and developed into a strategic concept of operation for a theater campaign or OPLAN. Otherwise, the CONOPS will be developed as a result of the COA selected by the JFC during COA development. The concept should be stated in terms of who, what, where, when, why, and how. It also contains the JFC's strategic vision, intent, and guidance for force projection operations, including mobilization, deployment, employment, sustainment, and redeployment of all participating forces, activities, and agencies.) (Refer to Annex C.)

(1) **Commander's Intent.** (This should describe the JFC's intent [purpose and end state], overall and by phase. This statement deals primarily with the military conditions that lead to mission accomplishment, so the commander may highlight selected objectives and their supporting effects. It may also include how the posture of forces at the end state facilitates transition to future operations. It may also include the JFC's assessment of the enemy commander's intent and an assessment of where and how much risk is acceptable during the operation. The commander's intent, though, is not a summary of the CONOPS.)

(a) **Purpose and End State.** *(See Chapter II, "Strategic Guidance and Coordination," for details on determining the end state.)*

(b) **Objectives.**

(c) **Effects,** if discussed.

(2) **General.** (Base the CONOPS on the JFC's selected COA. The CONOPS states how the commander plans to accomplish the mission, including the forces involved, the phasing of operations, the general nature and purpose of operations to be conducted, and the interrelated or cross-Service support. For a CCDR's contingency plan, the CONOPS should include a statement concerning the perceived need for Reserve Component mobilization based on plan force deployment timing and Reserve Component force size requirements. The CONOPS should be sufficiently developed to include an estimate of the level and duration of conflict to provide supporting and subordinate commanders a basis for preparing adequate supporting plans. To the extent possible, the CONOPS should incorporate the following:)

(a) JFC's military objectives, supporting desired effects, and operational focus.

(b) Orientation on the enemy's strategic and operational COGs.

(c) Protection of friendly strategic and operational COGs.

(d) Phasing of operations, to include the commander's intent for each phase.

1. **Phase I**

a. JFC's intent.

b. Timing.

c. Objectives and desired effects.

d. Risk.

e. Execution.

f. Employment.

(1) Land Forces.

(2) Air Forces.

(3) Maritime Forces.

(4) Space Forces.

(5) Cyberspace Forces.

(6) SOF.

g. **Operational Fires.** List those significant fires considerations on which the plan is based. The fires discussion should reflect the JFC's concept for application of available fires assets. Guidance for joint fires may address the following:

(1) Joint force policies, procedures, and planning cycles.

(2) Joint fire support assets for planning purposes.

(3) Priorities for employing target acquisition assets.

(4) Areas that require joint fires to support operational maneuver.

(5) Anticipated joint fire support requirements.

(6) Fire support coordination measures (if required).

See JP 3-09, Joint Fire Support, *for a detailed discussion.*

2. **Phases II through XX.** *(Cite information as stated in subparagraph 3a(2)(d)1 above for each subsequent phase based on expected sequencing, changes, or new opportunities.)*

b. **Tasks.** (List the tasks assigned to each element of the supported and supporting commands in separate subparagraphs. Each task should be a concise statement of a mission to be performed either in future planning for the operation or on execution of the OPORD.

The task assignment should encompass all key actions that subordinate and supporting elements must perform to fulfill the CONOPS, including operational and tactical deception. If the actions cannot stand alone without exposing the deception, they must be published separately to receive special handling.)

c. **Coordinating Instructions.** (Provide instructions necessary for coordination and synchronization of the joint operation that apply to two or more elements of the command. Explain terms pertaining to the timing of execution and deployments. Coordinating instructions should also include CCIRs and associated reporting procedures that may be expanded upon in Annex B [Intelligence], Annex C [Operations], and Annex R [Reports].)

4. **Administration and Logistics**

a. **Concept of Sustainment.** (This should provide broad guidance for the theater strategic sustainment concept for the campaign or operation, with information and instructions broken down by phases. It should cover functional areas of logistics, transportation, personnel policies, and administration.)

b. **Logistics.** (This paragraph addresses the CCDR's logistics priorities and intent: basing, combat, general, and geospatial engineering requirements, HNS, required contracted support, environmental considerations, mortuary affairs, and Service responsibilities. Identify the priority and movement of logistic support for each option and phase of the concept.)

c. **Personnel.** (Identify detailed planning requirements and subordinate taskings. Assign tasks for establishing and operating joint personnel facilities, managing accurate and timely personnel accountability and strength reporting, and making provisions for staffing them. Discuss the administrative management of participating personnel, the reconstitution of forces, command replacement and rotation policies, and required joint individual augmentation [JIA] to command headquarters and other operational requirements.) Refer to Annex E (if published).

d. **Public Affairs.** Refer to Annex F.

e. **Civil–Military Operations.** Refer to Annex G.

f. **Meteorological and Oceanographic Services.** Refer to Annex H.

g. **Environmental Considerations.** Refer to Annex L. See JP 3-34, *Joint Engineer Operations*.

h. **Geospatial Information and Services.** Refer to Annex B.

i. **Health Service Support.** Refer to Annex Q. (Identify planning requirements and subordinate taskings for joint health services functional areas. Address critical medical supplies and resources. Assign tasks for establishing joint medical assumptions and include them in a subparagraph.)

5. Command and Control

a. Command

(1) **Command Relationships.** (State the organizational structure expected to exist during plan implementation. Indicate any changes to major C2 organizations and the time of expected shift. Identify all command arrangement agreements and memorandums of understanding used and those that require development.)

(2) **Command Posts.** (List the designations and locations of each major headquarters involved in execution. When headquarters are to be deployed or the plan provides for the relocation of headquarters to an alternate command post, indicate the location and time of opening and closing each headquarters.)

(3) **Succession to Command.** (Designate in order of succession the commanders responsible for assuming command of the operation in specific circumstances.)

b. **Joint Communications System Support.** (Provide a general statement concerning the scope of communications systems and procedures required to support the operation. Highlight any communications systems or procedures requiring special emphasis.) Refer to Annex K.

[Signature]
[Name]
[Rank/Service]
Commander

Annexes:

A—Task Organization

B—Intelligence

C—Operations

D—Logistics

E—Personnel

F—Public Affairs

G—Civil-Military Operations

H—Meteorological and Oceanographic Operations

J—Command Relationships

K—Communications Systems

L—Environmental Considerations

M—Not currently used

N—Not currently used

P—Host-Nation Support

Q—Medical Services

R—Reports

S—Special Technical Operations

T—Consequence Management

U—Notional Counterproliferation Decision Guide

V—Interagency Coordination

W—Operational Contract Support

X—Execution Checklist

Y—Communication Synchronization

Z—Distribution

Note: Annexes A—D, K, and Y are required annexes for a crisis OPORD per APEX. All others may either be required by the JSCP or deemed necessary by the supported commander.

Intentionally Blank

APPENDIX B
STRATEGIC ESTIMATE

SECTION A. INTRODUCTION

1. Background

a. The strategic estimate is an analytical tool available to CCDRs before developing theater or functional strategies; theater, functional or DOD-wide campaign plans, subordinate campaign plans; and OPLANs. Strategic estimates provide the commander's perspective of the strategic and operational levels of the OE, threats and opportunities that could facilitate or hinder the achievement of GEF-directed objectives, desired changes to meet specified regional or functional objectives, and the commander's visualization of how those objectives might be achieved. Developed annually and regularly updated, the strategic estimate is the basis for developing the CCDR's theater or functional strategy.

b. The CCDR, the CCMD staff, supporting commands, and agencies assess the broad strategic factors that influence OE, thus informing the ends, ways, means, and risks involved in accomplishing the prescribed campaign objectives.

c. Both supported and supporting CCDRs prepare strategic estimates based on assigned tasks. CCDRs who support multiple commands may prepare strategic estimates for each supporting operation.

d. Section B, "Notional Strategic Estimate Format," presents a format a CCMD staff can use as a guide when developing a strategic estimate. The J-5 may provide the lead staff organization for the conduct of the strategic estimate with significant participation from the other staff directorates. The exact format and level of detail may vary somewhat among commands, based on theater-specific requirements and other factors.

e. The result of the strategic estimate is a better understanding and visualization of the complete OE to include adversaries, friends, and neutrals. The strategic estimate process is dynamic and continuous, and provides input for developing theater strategies and campaign plans. This strategic estimate is also the starting point for conducting more detailed staff estimates as well as the commander's estimate of the situation for a potential contingency.

f. The CCDRs strategic estimate should identify potential for spillover, both from the AOR or functional area perspective into other CCDRs' AORs or functional areas and into the CCDR's AOR or functional area based on operations and activities outside the AOR.

SECTION B. NOTIONAL STRATEGIC ESTIMATE FORMAT

2. Strategic Direction

(This section analyzes broad policy, strategic guidance, and authoritative direction to the theater or global situation and identifies strategic requirements in global and regional dimensions.)

a. **US Policy Goals.** (Identify the US national security or military objectives and strategic tasks assigned to or coordinated by the CCMD.)

b. **Non-US/Multinational Policy Goals.** (Identify the multinational [alliance or coalition] security or military objectives and strategic tasks that may also be assigned to, or coordinated by the CCMD.)

c. **Opposition Policy Goals and Desired End State**

d. **End State(s).** (Describe the campaign or operation objective[s] or end state[s] and related military objectives to achieve and end states to attain and maintain.)

3. **Operational Environment**

a. **AOR.** (Provide a visualization of the relevant geographic, political, economic, social, demographic, historic, and cultural factors in the AOR assigned to the CCDR.)

b. **Area of Interest.** (Describe the area of interest to the commander, including the area of influence and adjacent areas and extending into adversary territory. This area also includes areas occupied by enemy forces that could jeopardize the accomplishment of the mission.)

c. **Adversary Forces.** (Identify all states, groups, or organizations expected to be hostile to, or that may threaten, US and partner nation interests, and appraise their general objectives, motivations, and capabilities. Provide the information essential for a clear understanding of the magnitude of the potential threat.)

d. **Friendly Forces.** (Identify all relevant friendly states, forces, and organizations. These include assigned US forces, regional allies, and anticipated multinational partners. Describe the capabilities of the other instruments of power [diplomatic, economic, and informational], US military supporting commands, and other agencies that could have a direct and significant influence on the operations in this AOR.)

e. **Neutral Forces.** (Identify all other relevant states, groups, or organizations in the AOR and determine their general objectives, motivations, and capabilities. Provide the information essential for a clear understanding of their motivations and how they may impact US and friendly multinational operations.)

4. **Assessment of the Major Strategic and Operational Challenges**

a. This is a continuous appreciation of the major challenges in the AOR with which the CCDR may be tasked to deal.

b. These may include a wide range of challenges, from direct military confrontation, peace operations, and security cooperation (including building partner capacity and capability), to providing response to atrocities, humanitarian assistance, disaster relief, and stability activities.

5. Potential Opportunities

a. This is an analysis of known or anticipated circumstances, as well as emerging situations, that the CCMD may use as positive leverage to improve the theater strategic situation and further US or partner nation interests.

b. Each potential opportunity must be carefully appraised with respect to existing strategic guidance and operational limitations.

6. Assessment of Risks

Risk is the probability and consequence of loss linked to hazards.

a. This assessment matches a list of the potential challenges with anticipated capabilities in the OE.

b. Risks associated with each major challenge should be analyzed separately and categorized according to significance or likelihood (most dangerous or most likely).

c. The CCMD staff should develop a list of possible mitigation measures to these risks.

For more information on risk assessment, refer to CJCSM 3105.01, Joint Risk Analysis.

Intentionally Blank

APPENDIX C
STAFF ESTIMATES

SECTION A. INTRODUCTION

1. Role of Estimates

a. Staff estimates are central to formulating and updating military action to meet the requirements of any situation. Staff estimates should start with the strategic estimate and be comprehensive and continuous and visualize the future, while optimizing the limited time available to not become overly time-consuming. Comprehensive estimates consider both the quantifiable and the intangible aspects of military operations. They translate friendly and enemy strengths, weapons systems, training, morale, and leadership into combat capabilities. The estimate process requires the ability to visualize the battle or crisis situations requiring military forces.

b. Estimates are an essential part of the operational design process. Through their estimates, the staff provides expert assessment of the OE and relevant factors affecting effective planning and execution toward achievement of objectives and attainment of end states.

c. Estimates must be as thorough as time and circumstances permit. The JFC and staff must constantly collect, process, and evaluate information. They update their estimates:

(1) When the commander and staff recognize new facts.

(2) When they replace assumptions with facts or find their assumptions invalid.

(3) When they receive changes to strategic direction based on high-level civilian-military dialogue or when assessment recommendations are accepted to refine, adapt, or terminate.

d. Estimates for the plan in execution can often provide a basis for estimates for future plans, as well as changes to the plan in execution. Technological advances and near-real-time information estimates ensure that estimates can be continuously updated. Estimates must visualize the future and support the commander's visualization. They are the link between planning and execution and support continuous assessment. The commander's vision articulated in the strategic estimate directs the end state. Each subordinate unit commander must also possess the ability to envision the organization's desired end state, as well as those desired by their opposition counterpart. Estimates contribute to this vision. Failure to make staff estimates can lead to errors and omissions when developing, analyzing, and comparing COAs.

e. Not every situation will allow or require an extensive and lengthy planning effort. It is conceivable that a commander could review the assigned task, receive oral briefings, make a quick decision, and direct writing of the plan to commence. This would complete the process and might be suitable if the task were simple and straightforward.

f. Most commanders, however, are more likely to demand a thorough, well-coordinated plan that requires a complex staff estimate process. Written staff estimates are carefully prepared, coordinated, and fully documented.

g. Because of the unique talents of each directorate, involvement of all is vital. Each staff estimate takes on a different focus that identifies certain assumptions, detailed aspects of the COAs, and potential deficiencies that are simply not known at any other level, but nevertheless must be considered. Such a detailed study of the COAs involves the corresponding staffs of subordinate and supporting commands.

h. Each staff directorate:

(1) Reviews the OE, mission, and situation from its own staff functional perspective.

(2) Examines the factors and assumptions for which it is the responsible staff.

(3) Analyzes each COA from its staff functional perspective.

(4) Concludes whether the mission can be supported.

i. **The products of this process are revised, documented staff estimates.** These are extremely useful to the commander's J-5 staff, which extracts information from them for the commander's estimate. The estimates are also valuable to planners in subordinate and supporting commands as they prepare supporting plans. Although documenting the staff estimates can be delayed until after the preparation of the commander's estimate, they should be sent to subordinate and supporting commanders in time to help them prepare annexes for their supporting plans.

j. The principal elements of the staff estimates normally include mission, situation and considerations, analysis of opposing COAs, comparison of friendly COAs, and conclusions. The coordinating staff and each staff principal develop facts, assessments, and information that relate to their functional field. Types of estimates generally include, but are not limited to, operations, personnel, intelligence, logistics, communications, civil-military operations, military deception, and special staff. The details in each basic category vary with the staff performing the analysis. The principal staff directorates have a similar perspective—they focus on friendly COAs and their supportability. The J-2 staff estimate is separate from the intelligence estimate provided at the beginning of the planning process. The staff estimate is completed during the strategic guidance planning function and identifies available CCMD intelligence collection and analytic capabilities and anticipated shortfalls that may limit the J-2's ability to support the proposed friendly COAs. Also during the strategic guidance planning function, based on continuous JIPOE, the J-2 produces the intelligence estimate that serves as the baseline assessment of the OE, adversary capabilities (including requirements, vulnerabilities, and COGs), and an analysis of the various COAs available to the adversary according to its capabilities. The intelligence estimate conclusion will indicate the adversary's most likely COA, identify the effects of that COA on the accomplishment of the assigned mission, and where applicable, list exploitable adversary vulnerabilities associated with that COA. The intelligence estimate informs the commander's estimate.

k. In many cases, the activities in the JPP COA development step are not separate and distinct, as the evolution of the refined COA illustrates. Staff estimates and assumptions used in the initial COA development may be based on limited staff support. But as concept development progresses, COAs are refined and evolve to include many of the following considerations:

(1) What military operations are considered?

(2) Where they will be performed?

(3) Who will conduct the operation?

(4) When is the operation planned to occur?

(5) How will the operation be conducted?

l. An iterative process of modifying, adding to, and deleting from the original tentative list is used to develop these refined COAs. The staff continually evaluates the situation as the planning process continues. Early staff estimates are frequently given as oral briefings to the rest of the staff. In the beginning, they tend to emphasize information collection more than analysis. It is only in the later stages of the process that the staff estimates are expected to indicate which COAs can be best supported.

m. **Sample Estimate Format.** The following is a sample format that can be used as a guide when developing an estimate. The exact format and level of detail may vary somewhat among joint commands and primary staff sections based on theater-specific requirements and other factors. Refer to the CJCSM 3130.03, *Adaptive Planning and Execution (APEX) Planning Formats and Guidance,* for the specific format when there is a requirement for the supported JFC to submit a commander's estimate.

SECTION B. SAMPLE ESTIMATE FORMAT

2. Mission

a. **Mission Analysis**

(1) Determine the higher command's purpose. Analyze national security and national military strategic direction, as well as appropriate guidance in partner nations' directions, including long- and short-term objectives. Determine if a clearly defined military end state and related termination criteria are warranted.

(2) Determine specified, implied, and essential tasks and their priorities.

(3) Determine objectives and consider desired and undesired effects.

(4) Reassess if the strategic direction and guidance support the desired objectives or end state.

b. **Mission Statement**

(1) Express in terms of who, what (essential tasks), when, where, and why (purpose).

(2) Frame as a clear, concise statement of the essential tasks to be accomplished and the purpose to be achieved.

3. Situation and Courses of Action

a. **Situation Analysis**

(1) **Geostrategic Context**

(a) Domestic and international context: political and/or diplomatic long- and short-term causes of conflict; domestic influences, including public will, competing demands for resources and political, economic, legal, and moral constraints; and international interests (reinforcing or conflicting with US interests, including positions of parties neutral to the conflict), international law, positions of international organizations, and other competing or distracting international situations. Similar factors must be considered for theater and functional campaigns and noncombat operations.

(b) A systems perspective of the OE: all relevant political, military (see next paragraph), economic, social, information, infrastructure, and other relevant aspects. See Chapter IV, "Operational Art and Operational Design," for a discussion of developing a systems perspective.

(2) **Analysis of the Adversary/Competitors.** Scrutiny of the opponent situation, including capabilities and vulnerabilities (at the theater level, commanders normally will have available a formal intelligence estimate), should include the following:

(a) National and military intentions and objectives (to extent known).

(b) Broad military COAs being taken and available in the future.

(c) Military strategic and operational advantages and limitations.

(d) Possible external military support.

(e) COGs (strategic and operational) and decisive points.

(f) Specific operational characteristics such as strength, composition, location, and disposition; reinforcements; logistics; time and space factors (including basing utilized and available); and combat/noncombat efficiency and proficiency in joint operations.

(g) Reactions of third parties/competitors in theater and functional campaigns.

(3) **Friendly Situation.** Should follow the same pattern used for the analysis of the adversary. At the theater level, CCDRs normally will have available specific supporting estimates, including personnel, logistics, and communications estimates. Multinational operations require specific analysis of partner nations' objectives, capabilities, and vulnerabilities. Interagency coordination required for the achievement of objectives should also be considered.

(4) **Operational Limitations.** Actions either required or prohibited by higher authority, such as constraints or restraints, and other restrictions that limit the commander's freedom of action, such as diplomatic agreements, political or economic conditions in affected countries, and HN issues.

(5) **Assumptions.** Assumptions are intrinsically important factors upon which the conduct of the operation is based and must be noted as such. Assumptions should only be made when necessary to continue planning.

(6) **Deductions.** Deductions from the above analysis should yield estimates of relative combat power, including enemy capabilities that can affect mission accomplishment.

b. **COA Development and Analysis.** COAs are based on the above analysis and a creative determination of how the mission will be accomplished. Each COA must be adequate, feasible, and acceptable. State all practical COAs open to the commander that, if successful, will accomplish the mission. For a CCDR's strategic estimate, each COA typically will constitute an alternative theater strategic or operational concept and should outline the following:

(1) Major strategic and operational tasks to be accomplished in the order in which they are to be accomplished.

(2) Major forces or capabilities required (to include joint, interagency, and multinational).

(3) C2 concept.

(4) Sustainment concept.

(5) Deployment concept.

(6) Estimate of time required to achieve the objectives or termination criteria.

(7) Concept for establishing and maintaining a theater reserve.

4. Analysis of Adversary/Competitor Capabilities and Intentions

a. Determine the probable effect of possible adversary capabilities and intentions on the success of each friendly COA.

b. Conduct this analysis in an orderly manner by time phasing, geographic location, and functional event. Consider:

(1) The potential actions of subordinates two echelons down.

(2) Conflict termination issues; think through own action, opponent reaction, and counteraction.

(3) The potential impact on friendly desired effects and likelihood that the adversary's actions will cause specific undesired effects.

c. Conclude with revalidation of friendly COAs. Determine additional requirements, make required modifications, and list advantages and disadvantages of each adversary capability.

5. Comparison of Own Courses of Action

a. Evaluate the advantages and disadvantages of each COA.

b. Compare with respect to evaluation criteria.

(1) Fixed values for joint operations (the principles of joint operations, the fundamentals of joint warfare, and the elements of operational design).

(2) Other factors (for example, political constraints).

(3) Mission accomplishment.

c. If appropriate, merge elements of different COAs into one.

d. Identify risk specifically associated with the assumptions (i.e., what happens if each assumptions prove false).

6. Recommendation

Provide an assessment of which COAs are supportable, an analysis of the risk for each, and a concise statement of the recommended COA with its requirements.

APPENDIX D
OPERATION ASSESSMENT PLAN (EXAMPLES)

Intentionally Blank

ANNEX A TO APPENDIX D
OPERATION ASSESSMENT PLAN

1. Introduction

a. Operation assessment applies to both campaign plans and contingency plans and is continuous throughout planning and execution. For TCPs and FCPs, an assessment plan is prepared as an annex or appendix of the campaign plan as the campaign's operational approach is being developed and continues to be refined and adapted so long as the plan is in execution. The intermediate objectives and accompanying metrics are established that directly and measurably contribute to achieving campaign objectives. The campaign assessment plan is modified should a campaign branch contingency plan or crisis-generated order go into execution as a new campaign operation. For contingency plans, the supported CCDR determines whether an assessment plan is required to support the four planning functions in order to enhance effectiveness and keep it up to date and ready for transition to execution. As the contingency plan is modified to keep it effective and ready for transition to execution, the assessment plan, if required, may likewise need refinement and adaptation.

b. The impacts of friendly, adversary, and neutral actions in the OE to a military plan and its execution must be considered. Operation assessment can help to identify significant actions and evaluate the results of these actions. This typically requires collaboration with other agencies and multinational partners—preferably within a common, accepted process—in the interest of unified action and facilitating the commander's understanding of the OE. Intelligence collection and analysis, subordinate Service and functional components and JTFs, supporting commands and defense agencies, and country teams should report progress toward campaign objectives to the CCDR as specified in the campaign plan assessment annex or appendix.

c. Although there is no prescribed format for an assessment plan, the five paragraph APEX plan or order format is suggested as a template. The TCP or FCP assessment plan is included within the campaign plan as an annex or appendix. Contingency plan assessment plans, as required, may be an annex or appendix to the contingency plan or may be stand-alone plans. Tab A, "North Atlantic Treaty Organization Assessment Annex Sample Format," and Tab B, "United States Army Assessment Annex Sample Format," to this annex contain examples of assessment annex formats used by NATO and the US Army.

d. This appendix discusses the operation assessment plan development, steps 1 and 2 in operations assessment, covered in Chapter VI, "Operation Assessment."

2. Assessment Planning Steps

a. A common method for developing an assessment plan uses the six steps identified in Figure D-A-1.

b. **Step 1. Identify Information Requirements.** Strategic guidance documents such as the GEF and JSCP serve as the primary guidance to begin planning at the CCMDs. CCDRs

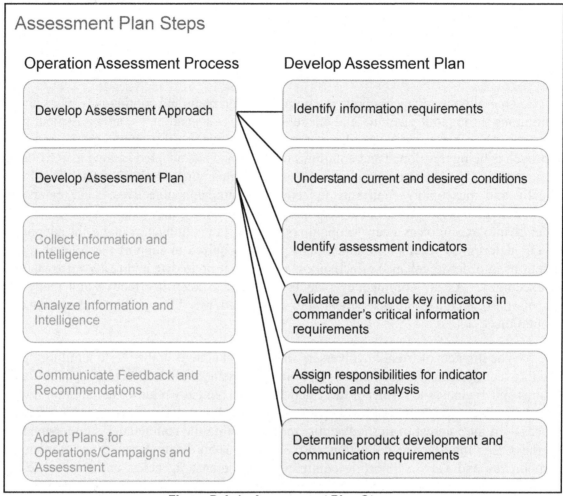

Figure D-A-1. Assessment Plan Steps

and other commanders may also initiate planning on their own authority when they identify a planning requirement not directed by higher authority. Subordinate components and commands typically begin planning based on higher headquarters guidance but should be aware of strategic guidance in order to properly nest supporting plans within plans being developed at higher headquarters. Military options are normally developed in combination with other nonmilitary options so the President can respond with the appropriate instruments of national power. Staffs begin updating their estimates and gather the information necessary for mission analysis and continued planning. Specific information gathered regarding assessment includes, but is not limited to:

(1) The higher headquarters' plan or order, including the assessment annex if available.

(2) If replacing a unit, any current assessment products.

(3) Relevant assessment products (classified or open source) produced by civilian, government, military, and partner nation organizations.

(4) The draft desired end state, objectives, and effects of the organization.

c. **Step 2. Understand Current and Desired Conditions**

(1) Fundamentally, operation assessment is about understanding current and desired conditions in the OE, observing changes in the OE, ascertaining the contribution of anticipated or completed tasks or missions to observed OE changes, and assessing progress or regression toward the desired OE conditions relative to the specified objective or end state. Staffs compare current conditions in the OA against the desired conditions. During mission analysis, JIPOE, and component-level intelligence preparation of the battlefield help develop an understanding of the current situation. The commander and staff identify the desired conditions and key underlying assumptions for an operation during joint planning. During execution, operational and intelligence reporting, the update of staff estimates, and any relevant information from external sources help them update and improve their understanding of the current conditions of the OE. Assumptions should be validated as soon as possible during execution. Likewise, desired conditions should be reevaluated as needed during execution.

(2) Understanding current and desired conditions requires acknowledging the underlying assumptions. Assumptions identified during planning are challenged during data analysis throughout operation assessment. If the assumptions are subsequently disproven, then reframing the problem may be appropriate.

(3) During initiation and operational design, commanders and selected key personnel develop and issue planning guidance that includes initial intent. That guidance is reviewed during mission analysis. Following mission analysis, commanders issue CCIRs, approve the mission statement, and issue additional guidance to guide the planning team during COA development. The end state in the initial commander's intent describes the conditions the commander wants to achieve. The staff section responsible for the assessment plan reviews each desired condition mentioned in the operational approach and commander's intent. These individual conditions provide focus for the overall planning, execution, and assessment of the operation. If the conditions that define the end state change during planning and execution, the staff updates these changes for the assessment plan.

(4) To assess progress, the staff identifies both the current situation and the desired end state. For example, the commander provides the end state condition "Essential services restored to pre-hostility levels." The staff identifies appropriate joint forces tasks, observable key indicators of task performance and effect(s) of task completion on OE conditions, and develops a plan to collect and analyze key indicator information while continuously monitoring OE conditions relative to this desired end state. These indicators also identify the current and pre-hostility levels of essential services across the OA. By taking these actions, the staff establishes a mechanism to assess progress toward these required conditions so that operations planned to achieve them are most effective.

d. **Step 3. Identify Assessment Indicators**

(1) An assessment plan should have a structure that begins with the operation or campaign's implied, specified, and essential tasks that, if successfully accomplished, should

achieve the campaign objectives or contingency end states. These tasks are used to establish measurable, achievable military objectives and accompanying metrics based on carefully selected MOEs and MOPs from among available indicators. Combined with continuous JIPOE, the MOEs and MOPs facilitate staff observations and analysis of changes in the OE and their impact on planning and execution.

(2) The assessment plan should focus on identifying those indicators and associated information and intelligence that accurately reflect changes in the OE. Analyses should identify whether desired conditions are being attained, and continually evaluate assumptions to validate or invalidate them. It should be noted that indicators and associated information and intelligence may require modification or replacement during planning and execution to respond to the dynamic conditions of the OE.

e. **Step 4. Assign Responsibilities for Collection and Analysis.** Indicator information is needed to help answer either an intelligence or information requirement. When the information required is unavailable from internal and external sources, these requirements can be integrated into the intelligence collection plan and tasked to intelligence collection assets. In any event, following collaboration with the affected organizations, responsibilities should be assigned for determining whether the needed information exists; and if it does not, responsibilities should then be assigned for collecting, processing, analyzing, and integrating required indicator information. In some cases, data may need to be collected from organizations external to the unit. For example, a HN's central bank may publish a consumer price index for that nation. The source for each indicator is identified in the assessment plan along with the staff element responsible for gathering it. Assessment information requirements compete with other information requirements for collection resources. When collection of data supporting an information requirement is not resourced, the staff will not have that information available for assessment, and will need to adjust the assessment plan accordingly.

See Annex B, "Data Collection Plan," to Appendix D, "Operation Assessment Plan (Examples)," for a discussion of DCP contents.

f. **Step 5. Assign Responsibilities for Analysis and Products.** In addition to gathering specific data, elements of the staff and subordinate and supporting organizations should be assigned to analyze indicator information and intelligence and develop analysis products and recommendations to decision makers. For example, the intelligence element leads the effort in assessing enemy forces and the engineering element leads the effort in assessing infrastructure development. The commander or designated representative should proactively require staff principals and subject matter experts to lead development of assessment products and communicate actionable recommendations synchronized with the operations cycle to support the commander's decisions.

g. **Step 6. Identify Communication Mechanisms.** An assessment product with meaningful recommendations that never reaches the appropriate decision maker wastes resources. The assessment plan should identify the best mechanisms (e.g., assessment reports, presentations, briefs, meetings) and frequency to communicate the findings and recommendations to decision makers. Considerations should include the commander's preferences and decision style, who else needs the information and recommendations (e.g., subordinate commanders, staff elements, external organizations), and the best way to disseminate the information. These mechanisms may include coordination requirements between staff elements and organizations, as well as follow up requirements and responsibilities for approved recommendations.

3. Assessment Plan Essentials

During the development of the assessment plan, the staff should:

a. Document the MOEs and MOPs in terms of acceptable conditions, rates of change, thresholds of success/failure, and technical/tactical triggers.

b. Document the selection of relevant aspects of the OE during mission analysis.

c. Document the development of information and intelligence requirements and record the linkage to key MOE and MOP indicators.

d. Document information and intelligence collection and analysis methods.

e. Establish methods to estimate risk integrated with the command's risk management process.

f. Establish methods to determine progress toward the desired end state.

g. Establish a method to evaluate triggers to the commander's decision points.

h. Develop a terms-of-reference document.

i. Establish the format for assessment products.

j. Coordinate development of recommendations.

Tab A, "North Atlantic Treaty Organization Assessment Annex Sample Format," and Tab B, "United States Army Assessment Annex Sample Format," provide sample formats of assessment annexes identified in NATO and US Army publications.

Intentionally Blank

TAB A TO ANNEX A TO APPENDIX D
NORTH ATLANTIC TREATY ORGANIZATION ASSESSMENT ANNEX
SAMPLE FORMAT

(Excerpt adapted from draft NATO Operations Assessment Handbook, Version 3.)

1. Introduction

The success of operations assessment will be predicated on the clear and concise orders set out in the operational plan prior to execution of an operation. ANNEX OO to the operational plan is reserved for the use of operations assessment (see NATO Comprehensive Operations Planning Directive). This chapter provides general guidance on the information that should be published in any given ANNEX OO.

2. Annex OO Template

The format of ANNEX OO should follow the guidance as given in the Comprehensive Operations Planning Directive, using the NATO standard six-paragraph format: Situation, Mission, CONOPS, Execution, Service and Support, and Command and Signal. The following template serves as a handrail for staff officers to ensure an effective Operations Assessment Annex to an OPLAN, OPORD, or CONPLAN. It provides suggested headings and recommended information for inclusion. At a minimum, all headings in the ANNEX should be published at the same time as the main body of the plan. (It is likely that the assessment plan will expand and refine over time and should be updated through the FRAGORD process accordingly.)

ANNEX OO OPERATION ASSESSMENT

1. SITUATION

a. **General.** Introduction to operation assessment, its purpose within the headquarters, relationship to the plan and the key references used in the design of the operation assessment plan.

b. **Purpose.** The purpose of the ANNEX.

2. MISSION. A clear, concise statement which states the operation assessment mission, with a clear purpose in support of commander's decision making.

3. CONCEPT OF OPERATIONS

a. **General CONOPS.** The general overview of the operation assessment will be described including the MOEs/MOPs, data collection, how the data will be analyzed to develop outputs, where the assessment will be used and what decisions it will support. Include reference to how lessons learned will be captured and the operation assessment refined.

b. **Operation Assessment Model/Process.** *A schematic drawing representing an overview of the process of operation assessment within the command.*

c. **Operation Assessment Results.** *How will the assessment products be presented? Where and who will use the output from the operation assessment?*

d. **DCP.** *Reference to how data will be collected using the data collection matrix detailed in Appendix I.*

4. EXECUTION

a. **Operations Assessment Battle Rhythm.** How the operations assessment will be executed with a battle rhythm and its relationship with the wider headquarters battle rhythm.

b. **Coordinating Instructions**

i. **Subordinate Command Tasks.** Tasks or responsibilities for subordinate Commands.

ii. **Supporting Command Tasks.** Tasks or responsibilities for supporting Commands.

iii. **HN Requests.** Requests to the HN for support. Identify overlaps with HN assessment capabilities.

iv. **Civilian Organizations Requests.** Requests to civilian organizations for support. Identify overlaps with civilian organizations assessment capabilities.

c. **Use of Tools for Operations Planning Functional Area Services** (commonly referred to as **TOPFAS) or other Operation Assessment-Related Software.** How the assessment will be executed using software applications, including databases and assessment tools such as TOPFAS.

5. SERVICE SUPPORT

a. **Contracting Support.** If any service contracts are to be established related to operations assessment, for example polling; detail plans for contracting here.

6. COMMAND AND SIGNAL

a. **C2.** Describe the relationship with other assessment cells.

b. **Liaison and Coordination.** Describe how to deal with issues and who the key points of contact are within the command.

c. **Reporting and Timing.** Provide key reports and timing for submission.

SIGNATURE BLOCK

(Excerpt adapted from Field Manual [FM] 6-0, *Commander and Staff Organization and Operations,* May 2014.)

ANNEX M (OPERATION ASSESSMENT) FORMAT AND INSTRUCTIONS

1. This annex provides fundamental considerations, formats, and instructions for developing Annex M (Assessment) to the BPLAN or order. This annex uses the five paragraph attachment format.

2. Commanders and staffs use Annex M (Assessment) as a means to quantify and qualify mission success or task accomplishment. The following staff entities are responsible for the development of Annex M (Assessment) at their various levels:

   a. Assistant COS, operations (G-3).

   b. Battalion or brigade operations staff officer (Army; Marine Corps battalion or regiment [S-3]).

   c. Assistant COS, plans (G-5) battalion or brigade.

   d. Civil affairs staff officer (Army; Marine Corps battalion or regiment [S-9]).

3. This annex describes the assessment concept of support objectives. This annex includes a discussion of the overall assessment concept of support, with the specific details in element subparagraphs and attachments.

SAMPLE FORMAT:

ANNEX M (ASSESSMENT) TO OPERATION PLAN/ORDER [number] [(code name)]—[issuing headquarters] [(classification of title)]

References: List documents essential to understanding the attachment.

   a. List maps and charts first. Map entries include series number, country, sheet names or numbers, edition, and scale.

   b. List other references in subparagraphs labeled as shown. List available assessment products that are produced external to this unit. This includes classified and open-source assessment products of the higher headquarters, adjacent units, key government organizations (such as the DOS), and any other relevant military or civilian organizations.

   c. Doctrinal references for assessment include Army Doctrine Reference Publication (ADRP) 5-0, *The Operations Process,* and FM 6-0.

Time Zone Used Throughout the Plan/Order: Write the time zone established in the BPLAN or order.

1. Situation. See the base order or use the following subparagraphs. Include information affecting assessment that paragraph 1 of the OPLAN or OPORD does not cover or that needs expansion.

a. **Area of Interest.** Describe the area of interest as it relates to assessment. Refer to Annex B (Intelligence) as required.

b. **Area of Operations.** Refer to Appendix 2 (Operation Overlay) to Annex C (Operations).

(1) **Terrain.** Describe the aspects of terrain that impact assessment. Refer to Annex B (Intelligence) as required.

(2) **Weather.** Describe the aspects of weather that impact assessment. Refer to Annex B (Intelligence) as required.

c. **Enemy Forces.** List known and templated locations and activities of enemy assessment units for one echelon up and two echelons down. List enemy maneuver and other area capabilities that will impact friendly operations. State expected enemy COAs and employment of enemy assessment assets. Refer to Annex B (Intelligence) as required.

d. **Friendly Forces.** Outline the higher headquarters' assessment plan. List designation, location, and outline of plans of higher, adjacent, and other assessment organizations and assets that support or impact the issuing headquarters or require coordination and additional support.

e. **Interagency, International, and NGOs.** Identify and describe other organizations in the area of operations that may impact assessment. Refer to Annex V (Interagency Coordination) as required.

f. **Civil Considerations.** Describe the aspects of the civil situation that impact assessment. Refer to Annex B (Intelligence) and Annex G (Civil-Military Operations) as required.

g. **Attachments and Detachments.** List units attached or detached only as necessary to clarify task organization. Refer to Annex A (Task Organization) as required.

h. **Assumptions.** List any assessment-specific assumptions that support the annex development.

2. Mission. State the mission of assessment in support of the BPLAN or order.

3. Execution

a. **Scheme of Operation Assessment.** State the overall concept for assessing the operation. Include priorities of assessment, quantitative and qualitative indicators, and the general concept for the way in which the recommendations produced by the assessment process will reach decision makers at the relevant time and place.

(1) **Nesting with Higher Headquarters.** Provide the concept of nesting of unit operation assessment practices with lateral and higher headquarters (include military and interagency organizations, where applicable). Use *Appendix 1 (Nesting of Operation Assessment Efforts) to Annex M (Assessment)* to provide a diagram or matrix that depicts the nesting of headquarters assessment procedures.

(2) **Information Requirements (DCP).** Information requirements for assessment are synchronized through the information collection process and may be CCIRs. Provide a narrative that describes the plan to collect the data needed to inform the status on metrics and indicators developed. The DCP should include a consideration to minimize impact on subordinate unit operations. Provide diagrams or matrices that depict the hierarchy of assessment objectives with the underlying MOEs, MOPs, indicators, and metrics. Provide MOEs with the underlying data collection requirements and responsible agency for collecting the data.

(3) **Battle Rhythm.** Establish the sequence of regularly occurring assessment activities. Explicitly state frequency of data collection for each data element. Include requirements to higher units, synchronization with lateral units, and products provided to subordinate units.

(4) **Reframing Criteria.** Identify key assumptions, events, or conditions that staffs will periodically assess to refine understanding of the existing problem and, if appropriate, trigger a reframe.

b. **Tasks to Subordinate Units.** Identify the unit, agency, or staff section assigned responsibility for collecting data, conducting analysis, and generating recommendations for each condition or MOE. Refer to paragraph 3a(2) (Information Requirements) of this annex as necessary.

c. **Coordinating Instructions.** List only instructions applicable to two or more subordinate units not covered in the BPLAN or order. Use Appendix 3 (Assessment Working Group) to Annex M (Assessment) to include quad charts that provide details about meeting location, proponency, members, agenda, and inputs or outputs.

4. Sustainment. Identify priorities of sustainment assessment key tasks and specify additional instructions as required. Refer to Annex F (Sustainment) as required.

a. **Logistics.** Identify unique sustainment requirements, procedures, and guidance to support assessment teams. Use subparagraphs to identify priorities and specific instructions for assessment logistics support. Refer to Annex F (Sustainment) and Annex P (Host-Nation Support) as required.

b. **Personnel.** Use subparagraphs to identify priorities and specific instructions for human resources support, financial management, legal support, and religious support. Refer to Annex F (Sustainment) as required.

c. **Health Services.** Identify availability, priorities, and instructions for medical care. Refer to Annex F (Sustainment) as required.

5. Command and Signal

a. **Command.** State the location of key assessment cells. State assessment liaison requirements not covered in the unit's SOPs.

(1) **Location of the Commander and Key Leaders.** State the location of the commander and key assessment leaders.

(2) **Succession of Command.** State the succession of command, if not covered in the unit's SOPs.

(3) **Liaison Requirements.** State the assessment liaison requirements not covered in the unit's SOPs.

b. **Control**

(1) **Command Posts.** Describe the employment of assessment-specific command posts, including the location of each command post and its time of opening and closing.

(2) **Reports.** List assessment-specific reports not covered in SOPs. Refer to Annex R (Reports), as required.

Annex H (Signal), as required.

OFFICIAL:

ACKNOWLEDGE: Include only if attachment is distributed separately from the base order. [Commander's last name] [Commander's rank] The commander or authorized representative signs the original copy of the attachment. If the representative signs the original, add the phrase "For the Commander." The signed copy is the historical copy and remains in the headquarters' files.

[Authenticator's name]

[Authenticator's position]

Use only if the commander does not sign the original attachment. If the commander signs the original, no further authentication is required. If the commander does not sign, the signature of the preparing staff officer requires authentication and only the last name and rank of the commander appear in the signature block.

ATTACHMENTS: List lower-level attachment (appendices, tabs, and exhibits).

Appendix 1—Nesting of Operation Assessment Efforts

Appendix 2—Framework for Assessment Framework

Appendix 3—Assessment Working Group

DISTRIBUTION: Show only if distributed separately from the base order or higher level attachments.

Intentionally Blank

ANNEX B TO APPENDIX D
DATA COLLECTION PLAN

1. Developing the Data Collection Plan

a. After the assessment indicators have been established, the staff develops a DCP in coordination with planners. This process should include members of staff who will become responsible for collecting data. Although there is no set format for a DCP, it should, at a minimum, identify the following for each indicator:

    (1) Data parameters, such as:

        (a) Units of measurement.

        (b) Scale, if appropriate.

        (c) Categorization for nominal or interval data.

        (d) Upper and lower bounds (if required).

        (e) Additional criteria.

    (2) Source of the data.

    (3) Method of collection.

    (4) Party responsible for collection.

    (5) Format in which data should be recorded.

    (6) Required frequency of collection.

    (7) Data recipients (who needs the data).

    (8) Required frequency of reporting.

    (9) Additional information.

b. The creation of the data collection matrix assists in clarifying the 'measurability' of the selected indicator and may result in further refinement. The DCP should always be synchronized and deconflicted with established reports across the command.

    (1) Some data for indicators, particularly those associated with performance, may already be organic—generated, captured, and reported by units within the command structure—while some might be reported by external nonmilitary organizations. While some of this information may be available prior to execution, the majority of performance related reporting occurs following execution.

(2) Other data for indicators associated with impacts on the OE will require the designation of observers as part of the intelligence collection plan or the development of another mechanism for collection. Generally, it is helpful to establish a baseline as early as possible from which subsequent change can be determined. It may be possible to collect on some indicators prior to beginning an operation. In other cases, the operation will begin, and data are collected as early as feasible.

c. The DCP should be published with the final operation or campaign plan/order. Once the campaign or operation is approved by the commander, all levels of command should start the operation assessment collection process. Throughout planning and execution, the collection plan should be modified as required until the plan or operation is terminated.

2. Data Sources

a. Staff elements, in conjunction with commander's communication synchronization personnel, should identify expected and available sources of data. Data can originate from a variety of sources, including but not limited to:

(1) Local population (formal or informal surveys).

(2) HN officials (formal or informal surveys).

(3) HN records.

(4) Other government departments and agencies (i.e., embassies, development departments).

(5) International organizations working in area (e.g., UN, World Bank, International Monetary Fund, European Union, Organization for Economic Cooperation and Development).

(6) NGOs working in area.

(7) Friendly force observations (e.g., patrol reports, intelligence).

(8) Media and other open sources (e.g., local, national, and home radio, Internet, social media, television, and print sources).

(9) Commercial data sources (e.g., DataCards)

(10) Assessment products from superior, subordinate, and supporting commands

(11) Subject matter experts within the command.

(12) Lessons learned and historical records.

b. Each data source requires appropriate scrutiny prior to and during use. When classification rules allow, the source should always be linked to the information collected to help provide full disclosure when reporting. Without this information the credibility of any recommendation could be disputed if the analysis and communication products appear to be overly positive or negative when compared with general perceptions and expectations.

c. Some data sources may be using that data for their own assessment purposes. It should be considered whether data sourced from other organizations is raw or processed. In the case of processed data, knowledge about the raw data, assumptions, and processing methods involved should be obtained.

d. The staff should specify the expected source of data for each indicator and, if available, identify back-up or corroborating sources for the following reasons:

(1) **Data From Multiple Sources is More Easily Verified.** A data item from one source is not as valuable as when the same data item is corroborated by other sources.

(2) **Mitigate Human Bias.** If the data item involves visual observations (e.g., number of open shops) or perceptive observations (e.g., sense of security in the town), the data may vary significantly depending on the source chosen. Bias is a danger when using human subjects

(3) **Keep Track of Data and Its Origin.** Whether data is taken from one or multiple sources, source identity is important for analysis purposes.

(4) **Data Archiving.** Historical records and data backups are essential. In addition to capturing lessons learned, analysts can work to improve measures by performing trend analysis of data over time. Improved historical data capture can also improve the ability to use predictive analytical techniques where opportunities arise.

3. Methods of Collection

a. Throughout planning and execution, planners and staff should identify assessment-related data collection methods. They should identify resources required to achieve data collection, prepare data collection orders for subordinate and supporting commands, and identify appropriate liaison with non-military actors to set up data exchange procedures. The command's mindset should be: "Everyone is a collector." Those responsible for the assessment process should remember this when determining collection methods. Since collection resources are often limited, planners and staffs should seek to establish a balance between the resources used for data collection and resources for other military tasks.

b. Figure D-B-1 provides some examples of data collection methods and associated advantages and disadvantages.

4. Assign Responsibility for Data Collection

a. The staff should normally assign individual units or organizations with responsibility for each data collection item in the data collection matrix. In some situations, an individual could be assigned the responsibility for one or more indicators.

Data Collection Methods

Data Collection Method	Description	Advantages	Disadvantages
Military Survey	A selective and planned questioning by military forces.	• Ease of tasking military forces. • Superior mobility. • Ability to access difficult environments. • Good for gathering raw qualitative data.	• Response of subjects may be biased by negative perception of military. • Military forces may not have specific skills in surveying.
Survey	A selective and planned questioning of subjects by nonmilitary parties (e.g., charities, nongovernmental organizations, or specialists).	• Survey by independent bodies can be more impartial. • Reduced burden on military. • May use survey specialists.	• Difficult to task nonmilitary organizations and extra financial cost may be involved. • Nonmilitary organizations may have reduced ability to access difficult environments.
Focus Group	A group of people are asked about their perceptions, opinions, beliefs towards a subject of interest. Questions are asked in an interactive group setting where participants are free to talk with other participants.	• Enables collection of in depth attitudes, belief, and anecdotal data. • Group dynamics facilitate idea generation. • Participants not required to read or write, relies on oral communication.	• Requires strong, trained facilitator. • Difficult to make conclusions that represent a population view. • Limited ability to repeat data collection over time.
Structured Interview	A planned, targeted discussion with a subject where the objectives of the discussion are pre-determined and noted in the data collection matrix.	• Often captures richer information than a survey. • Many methods may be used: face to face, electronic mail, telephone, video conference, etc. • Provides opportunity to probe and explore ideas in depth.	• Interviews are very time consuming. • May require trained interviewer or transcriber.
Structured Observation/ Debrief	A set of specific observations collected during routine work, followed by formal question, or asked to report observations at a specific time.	• Good approach to discover behaviours.	• Structured observations can be time consuming. • Human bias of difference in perceptions.
Military Situation Report	A formatted report intended to convey a pre-defined set of information in relation to a specific event or activity, or a routine (time dependant) report.	• A standardised set of information which helps in consistency of reporting. • A normal part of military business rather than an extra burden.	• Limited opportunities for reporting nonstandard data, or for changing report formats for mission specific data.
Automatic Media Collection	An automated collection and analysis of open source media (e.g., Rich Site Summary feeds, online market data, social media, country watch reports) or closed-source media.	• Resources and time are saved in the efforts required for the data collection.	• More analyst time is required to sort through data. • Automatic methods may either collect too much or too little, or miss vital data.
Manual Media Collection	Manual observation (e.g., reading documents, logging events, photocopying of open-source media or closed-source media.	• The data collection can be more thorough, with a certain amount of analysis being done simultaneously.	• Manual data collection is time consuming.

Figure D-B-1. Data Collection Methods

b. Codifying responsibility is important for the following reasons:

(1) Assigning responsibility increases the likelihood of the task being accomplished.

(2) A reporting chain is clearly identified and communicated.

(3) In the event of a data query, the analyst can direct questions to the person, unit or organization responsible for that data item.

(4) Facilitates data archiving and data analysis.

(5) If data collection tasks fall to persons or organizations outside the military, a flag is raised for an action to establish links with those particular persons or organizations.

(6) If data originates from an organization not likely to be compliant, then a flag is raised to seek an alternative source of data or even an alternative indicator.

5. Collection and Reporting Timelines

Planners and staff, with guidance from the commander, should determine the frequencies for data collection and reporting.

a. **Data Collection Frequency. For each indicator,** the number of times (e.g., per day, week, month) that the data should be collected.

(1) Is the requested frequency commensurate with the possible observable change? It makes no sense to record the number of enemy aircraft sorties per day when only a few occur each month. Conversely, if the incidence of a highly contagious disease in a refugee camp was being monitored, daily figures would be more appropriate than the number of new infections each month.

(2) Is the data collection likely to be influenced by important events? For example, while 'attacks per month' is sufficient for most cases, in the month leading up to the regional elections it may be prudent to capture 'attacks per day.'

b. **Data Reporting Frequency.** The number of times per day, week, month, etc., that the data should be reported to identified users. Data may be reported to different users at different frequencies.

c. **Data collection frequency and data reporting frequency may not be the same.** Typically the requested collection frequency will support the most rapid reporting frequency. For example, the requirement to collect the "number of attacks" on a daily basis can support a tactical commander who requires reporting on the "attacks per day," but also support a higher command's requirement for 'attacks per week" or "attacks per month."

Intentionally Blank

1. Purpose

GFM allows SecDef to strategically manage the employment of the force among the CCDRs. This is accomplished via three related processes: assignment, allocation, and apportionment as depicted in Figure E-1. These processes allow SecDef to strategically manage US Armed Forces to accomplish priority missions assigned to the CCDRs enabling the DOD to meet the intent of the strategic guidance contained in the DSR, NMS, UCP, GEF, and Defense Planning Guidance. The assignment and allocation processes allow SecDef to distribute forces to the CCDRs in a resource-informed manner while assessing the risks to current operations and missions; potential future contingencies; and the health, readiness, and availability of the current and future force. Based on the number of forces employed globally, GFM advises CCDRs of the Services' capacity to deploy forces to CCDRs to meet potential future contingencies. GFM also assesses shortfalls for current and potential future contingencies, mitigates current shortfalls, and informs the Services' force development processes.

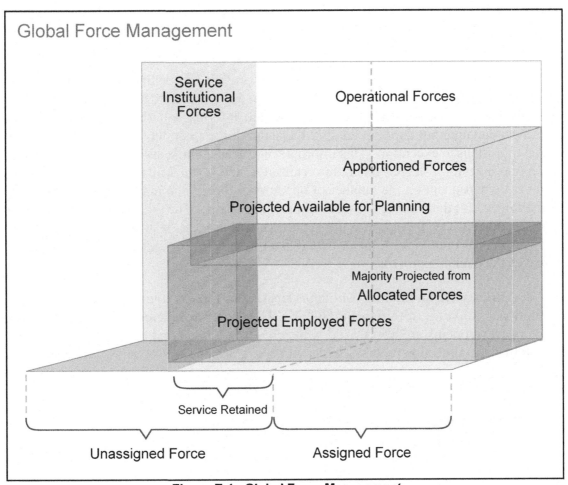

Figure E-1. Global Force Management

a. **Assignment.** Fulfills the Military Departments, Title 10, USC, Section 162, responsibility to assign specified forces to CCDRs or to the US Element, North American Aerospace Defense Command as directed by SecDef to perform missions assigned to those commands. CCDRs exercise combatant command (command authority) over forces assigned to them. Assignment of forces is conducted annually and documented in the GFMIG. This is published bi-annually on even years in the GFMIG and, in the years when the GFMIG is not updated, in a memorandum published separately.

b. **Allocation.** Pursuant to Title 10, USC, Section 162, "(3) A force assigned to a CCMD or the United States element of the North American Aerospace Defense Command under this section may be transferred from the command to which it is assigned only (A) by authority of SecDef; and (B) under procedures prescribed by the Secretary and approved by the President." Under this authority, SecDef allocates forces between CCDRs. The allocation process adjusts the distribution of forces among the CCDRs to meet force requirements in support of current operations and campaign plans to mitigate near-term military and strategic risk. SecDef decisions to allocate forces are published in the CJCS annual DEPORD called the GFMAP and its associated annexes. When transferring forces, SecDef will specify the command relationship the gaining CCDR will exercise and the losing CCDR will relinquish. CCDRs request joint individual augmentees, when required, to man a JTF headquarters. These JIA requirements are allocated by SecDef and ordered in the GFMAP. Further discussion of the GFM allocation process can be found in the GFMIG and CJCSM 3130.06, *(U) Global Force Management Allocation Policies and Procedures.*

c. **Apportionment.** Apportioned forces provide an estimate of the Military Departments' capacity to generate capabilities that can reasonably be expected to be made available along general timelines. This estimate informs and shapes CCDR resource informed planning, but does not identify the actual forces that may be allocated for use if a plan transitions to execution. This informs senior leadership's assessment of plans based on force inventory, force generation capacity, and availability. Apportionment is necessarily dependent on the number of operational forces, the readiness and availability of the forces, and the number of forces employed globally. The GEF and GFMIG provide strategic guidance with respect to the apportionment process. CJCSI 3110.01, *(U) 2015 Joint Strategic Capabilities Plan (JSCP)*, contains guidance for implementing apportioned forces in the planning process.

d. **Military Departments.** Military Department forces required to execute Service institutional activities specified in Title 10, USC, are considered "unassigned." The Military Departments are also tasked with providing trained and equipped forces to the CCDRs via the allocation process. These forces are designated as "Service retained."

2. **Authorities and Responsibilities**

a. **Strategic Guidance for GFM**

(1) **Title 10, USC.** Title 10, UCS, governs the US Armed Forces and provides for the organization of DOD, including the Military Departments and Reserve Component, and establishes statutory responsibilities and requirements.

(2) **UCP.** The UCP, approved by the President, provides direction to all CCDRs including their missions, responsibilities, and AOR.

(3) **GEF.** The GEF, approved by the President and SecDef, translates national security strategic objectives into a prioritized and comprehensive planning tool to guide the employment of US Armed Forces. It consolidates guidance for campaign planning, security cooperation, GDP, and GFM.

(4) **GFMIG.** The GFMIG, approved by SecDef, integrates force assignment, apportionment, and allocation processes to improve the DOD's ability to manage forces from a global perspective. It provides guidance and assigns responsibilities for performing the assignment, allocation and apportionment processes, and contains the Forces For Unified Commands ("Forces For") tables specifying the assignment of forces. In years that the GFMIG is not updated, the "Forces For" tables are published separately.

(5) **CJCSI 3110.01, *(U) 2015 Joint Strategic Capabilities Plan (JSCP).*** The JSCP, approved by the CJCS, provides policy to accomplish tasks and missions based on near-term military capabilities. It implements the strategic policy guidance provided in the GEF and initiates the planning process for the development of campaign and top-priority contingency plans. The JSCP contributes to the CJCS's statutory responsibilities to assist the President and SecDef in providing for the strategic direction to the US Armed Forces and further explains apportioned forces.

b. **GFM Stakeholders Responsibilities**

(1) **SecDef.** Title 10, USC, Section 131, authorizes SecDef to act as the principal assistant to the President in all matters relating to the DOD. SecDef is responsible for directing the OSD in the development of DOD guidance and policy. SecDef is the decision authority for GFM assignment and allocation.

(2) **CJCS.** Title 10, USC, Sections 152-153, authorizes the CJCS to act as the principal military adviser to the President, NSC, HSC, and SecDef. The CJCS heads the JCS but does not exercise military command. The CJCS is the decision authority of GFM apportionment. The CJCS issues orders implementing the President's or SecDef's direction. At the direction of the CJCS, the JS conducts the following activities in support of GFM:

(a) Per the GFMIG, the JS J-8 prepares and publishes the force assignment tables/"Forces For" and the apportionment tables.

(b) Per the GFMIG, the JS J-3 is responsible to the CJCS for leading and coordinating the GFM allocation process to validate and provide recommended sourcing solutions to meet CCDRs' force and JIA requirements that cannot be met with assigned forces. The JS J-3 is the staff lead for preparation and coordination of the GFMAP consolidating JFP allocation recommendations for SecDef decision. The JS J-3 also performs the duties of the JFP for conventional forces responsible for identifying and recommending contingency and execution sourcing solutions in coordination with the Secretaries of Military Departments, CCMDs, DOD agencies, other JFPs, and the JFM for all force and individual requirements. Individual augmentee requirements are not normally contingency sourced.

(3) **CCDRs.** Per Title 10, USC, Sections 161-168, a CCDR is responsible to the President and SecDef for the performance of missions assigned to their geographic or functional CCMD. CCDRs perform their duties under the authority, direction, and control of SecDef and provide authoritative direction to the subordinate commands and forces assigned to their respective CCMD. CCDRs have the authority to employ forces within their CCMDs to carry out missions assigned to the CCMD. CCDRs act as the supported commander for the execution of their assigned missions which involves the responsibility of synchronizing military engagement, security cooperation, and deterrence activities required to achieve the desired objectives. They may simultaneously be a supporting commander to another CCDR's mission. The GFM responsibilities of CCDRs include the following:

(a) **FP.** Per the GFMIG, CCDRs with assigned forces are designated FPs and will develop and provide force sourcing solutions, via the JS J-35 or one of the JFPs, in response to CCDR force requirements.

(b) **JFPs.** Per the UCP and GFMIG, Commander, United States Special Operations Command (USSOCOM), serves as the JFP for special operations forces. Commander, USTRANSCOM, serves as the JFP for mobility forces. Each is responsible for identifying and recommending force sourcing solutions in coordination with the Secretaries of Military Departments, CCDRs, DOD agencies, other FPs and JFPs, and the JFM for validated force requirements.

(c) **JFM.** Per the GFMIG, Commander USSTRATCOM will serve as the JFM for missile defense forces. In coordination with the JS J-35 and supported CCDRs, USSTRATCOM will collaboratively develop and provide sourcing recommendations for global missile defense force requirements.

(4) **Secretaries of Military Departments.** Title 10, USC, Section 111, specifies the Secretaries of the Military Departments as part of the executive structure of the DOD. They are subject to the direction and control of SecDef and responsible for the organization, development, and programing for their respective Military Departments, consistent with policy and national security objectives. The GFM responsibilities of the Military Department Secretaries include:

(a) Per the GFMIG, Military Departments coordinate directly with CCMDs, JFPs, and the JS J-35 to develop recommended global sourcing solutions. This activity is currently executed by the Military Departments through designated Service FP organizations.

(b) Assignment of forces under DOD jurisdiction, as directed by SecDef, to the unified and specified CCMDs or to the US Element, North American Aerospace Defense Command to perform missions assigned to those commands.

(c) Prepare and deploy trained and equipped Service-retained forces to the CCDRs to carry out missions assigned to the CCDRs.

(5) **United States Coast Guard (USCG).** The GFM responsibilities of USCG include:

(a) Per the GFMIG, USCG coordinates directly with CCMDs and the JS J-35 to develop recommended global sourcing solutions for USCG forces.

(b) Normally USCG forces are not assigned to CCMDs, but are apportioned for CCMD planning and allocated in the GFMAP through the GFM process.

(c) Prepare and deploy trained and equipped USCG forces to the CCDRs to carry out missions assigned to the CCDRs.

(6) **DOD Agencies.** Per the GFMIG, DOD agencies coordinate directly with CCMDs, JFPs, and the JS J-35 to develop recommended global sourcing solutions.

3. Global Demand

The demand for forces originates with the CCDRs as they require forces to execute their campaign plans, operations, exercises, and other military activities. To make risk informed decisions, SecDef must consider the entire demand on the force pool, priority of each military operation or activity, and impact on the readiness and availability of the remaining forces to respond. Global demand originates primarily from the CCDRs as follows:

a. **Campaign Plans.** During the execution of CCDR campaign plans, forces and joint individual augmentees are required to support the events and activities that comprise the functional and TCP. These force requirements may be met with assigned and/or allocated forces. CJCSM 3130.01, *Campaign Planning Procedures and Responsibilities,* details campaign planning.

b. **Contingency Operations.** CCDRs require forces and joint individual augmentees to execute operations. These force requirements may be met with forces assigned to the CCDR or forces and joint individual augmentees requested and allocated by SecDef in the GFMAP. Under crisis time constraints, allocation decisions of SecDef may be made by verbal orders of the commander. Verbal orders should always be followed up with a written order as soon as practicable.

c. **Rotational Force Planning.** Requirements for allocated forces and joint individual augmentees may be enduring for a period of time beyond a single deployment cycle. Supported CCDRs identify enduring force requirements and request them via the allocation process for SecDef approval. CJCSM 3130.06, *(U) Global Force Management*

Allocation Policies and Procedures, provides detailed guidance for rotational force planning.

d. **Joint Exercises.** Encompasses all CCDR requirements for CJCS-directed and CCDR high-priority exercises. Forces participating in joint exercises are not usually allocated by SecDef. Military Departments provide forces in support of CCDR joint exercises under Title 10, USC, authority to conduct training. Forces sourced to joint exercises may be subsequently allocated to meet a higher-priority operational force requirement. CJCSM 3500.03, *Joint Training Manual for the Armed Forces of the United States,* provides detailed procedures for sourcing forces for joint exercises.

e. **Potential Future Contingencies.** CCDR plans contain requirements for forces to respond to potential future contingencies. When developing these plans, CCDRs consider their assigned forces, apportioned forces, and the need to request allocated forces should a plan be executed. CCDRs also consider the need to request rotational forces should the plan be executed. The risks to these potential future contingencies are considered by the FPs, JFPs, JS, OSD, CJCS, and SecDef when execution sourcing forces to respond to current operations and campaign events.

4. Sourcing

The concept of strategy-driven and resource-informed planning requires the development of plans based on the near-term readiness of the force. GFM procedures allow proactive resource and risk-informed planning assumptions and estimates and execution decision making regarding US Armed Forces. Time-phased force requirements are documented in a TPFDD. Within GFM, there are three levels of matching forces to requirements, depending on the end state required:

a. **Preferred Force ID.** As a planning assumption, CCMD planners identify actual units as preferred forces necessary to continue planning and assess the feasibility of a plan. The number of identified preferred forces should be within the quantities of those force types apportioned. Preferred forces are planning assumptions only and do not indicate that those forces will be contingency or execution sourced. Although not preferred forces, CCMD planners can consider capabilities available in current contracts, diplomatic agreements, and task orders as available for planning.

b. **Contingency Sourcing.** Contingency sourcing is led by the JS J-35 and JFPs as a sourcing feasibility assessment. This is normally performed prior to the final IPR for plan approval and during the plan assessment process as part of the joint combat capability assessment (JCCA) during the APEX plan development and plan assessment planning functions. The JS J-5 provides specific guidance through a list of sourcing assumptions and planning factors contained in the contingency sourcing message. The resultant contingency sourced forces represent a snapshot in time sourcing feasibility of the plan for senior leaders. CJCSI 3401.01, *Joint Combat Capability Assessment,* details the JCCA process and CJSCM 3130.06, *(U) Global Force Management Allocation Policies and Procedures,* provides detailed contingency sourcing procedures.

c. **Execution Sourcing.** During execution, the supported CCDR may task their assigned forces to fill force requirements to perform authorized missions. These requirements constitute the assigned force demand. If additional forces are required, the supported CCDR requests those forces through the GFM allocation process for consideration by SecDef. The SecDef's decision to allocate forces involves weighing the FP's risks of sourcing with operational risks to both current operations and potential future contingencies. The SecDef's decisions are ordered in the GFMAP directing a FP to source the force. The FP identifies the unit and issues DEPORDs, via the chain of command to the unit or individual. CJCSM 3130.06, *(U) Global Force Management Allocation Policies and Procedures,* provides detailed execution sourcing procedures.

Intentionally Blank

APPENDIX F
FLEXIBLE DETERRENT OPTIONS AND FLEXIBLE RESPONSE OPTIONS

FDOs and FROs are executed on order and provide scalable options to respond to a crisis. Commanders include FDOs and FROs as part of their plans to provide adaptive military options for SecDef or the President to deter or respond to a crisis. Both provide the ability to scale up (escalate) or de-escalate based on continuous assessment of an adversary's actions and reaction. While FDOs are primarily intended to prevent the crisis from worsening and allow for de-escalation, FROs are generally punitive in nature.

SECTION A. FLEXIBLE DETERRENT OPTIONS

1. General

a. FDOs are preplanned, deterrence-oriented actions tailored to signal to and influence an adversary's actions. They are established to deter actions before or during a crisis. If necessary, FDOs may be used to prepare for future operations, recognizing they may well create a deterrent effect.

b. FDOs are developed for each instrument of national power—diplomatic, informational, military, and economic—but they are most effective when combined across the instruments of national power. FDOs facilitate early strategic decision making, rapid de-escalation, and crisis resolution by laying out a wide range of interrelated response paths.

c. FDOs provide options for decision makers during emerging crises to allow for gradual increase in pressure to avoid unintentionally provoking full-scale combat and to enable them to develop the situation and gain a better understanding of an adversary's capabilities and intentions. FDOs are elements of contingency plans executed to increase deterrence in addition to but outside the scope of the ongoing operations.

d. Examples of FDOs for each instrument of national power are listed in Figures F-1 through F-4. Key objectives of FDOs are:

(1) Communicate the strength of US commitments to treaty obligations and regional peace and stability.

(2) Confront the adversary with unacceptable costs for their possible aggression.

(3) Isolate the adversary from regional neighbors and attempt to split the adversary coalition.

(4) Rapidly improve the military balance of power in the AOR without precipitating armed response from the adversary.

(5) Develop the situation without provoking the adversary to better understand his capabilities and intentions.

Examples of Requested Diplomatic
Flexible Deterrent Options

- Alert and introduce special teams (e.g., public diplomacy).
- Reduce international diplomatic ties.
- Increase cultural group pressure.
- Promote democratic elections.
- Initiate noncombatant evacuation procedures.
- Identify the steps to peaceful resolution.
- Restrict activities of diplomatic missions.
- Prepare to withdraw or withdraw US embassy personnel.
- Take actions to gain support of allies and friends.
- Restrict travel of US citizens.
- Gain support through the United Nations.
- Demonstrate international resolve.

Figure F-1. Examples of Requested Diplomatic Flexible Deterrent Options

 e. Deterrence is perception based. The US must be certain the audiences to be deterred are aware of the actions.

Examples of Requested Informational
Flexible Deterrent Options

- Impose sanctions on communications systems and intelligence, surveillance, and reconnaissance (ISR) technology transfer.
- Protect friendly communications systems and intelligence collection assets (defensive space control, defensive cyberspace operations, operations security, cybersecurity).
- Increase public awareness of the problem and potential for conflict.
- Make public declarations of nonproliferation policy.
- Increase communication systems and ISR processing and transmission capability.
- Interrupt satellite downlink transmissions.
- Publicize violations of international law.
- Publicize increased force presence, joint exercises, military capability.
- Increase informational efforts:
  - Influence adversary decision makers (political, military, and social).
  - Promote mission awareness.
  - Increase measures directed at the opponent's military forces.
- Implement meaconing, interference, jamming, and intrusion of enemy informational assets.
- Maintain an open dialogue with the news media.
- Take steps to increase US public support.
- Ensure consistency with strategic guidance.

Figure F-2. Examples of Requested Informational Flexible Deterrent Options

Examples of Requested Military
Flexible Deterrent Options

- Increase readiness posture of in-place forces.
- Upgrade alert status.
- Increase intelligence, surveillance, and reconnaissance.
- Initiate or increase show-of-force actions.
- Increase training and exercise activities.
- Increase defense support to public diplomacy.
- Increase information operations.
- Deploy forces into or near the potential operational area.
- Increase active and passive protection measures.

Figure F-3. Examples of Requested Military Flexible Deterrent Options

**Examples of Requested Economic
Flexible Deterrent Options**

- Freeze or seize real property in the US where possible.
- Freeze monetary assets in the US where possible.
- Freeze international assets where possible.
- Encourage US and international financial institutions to restrict or terminate financial transactions.
- Encourage US and international corporations to restrict transactions.
- Embargo goods and services.
- Enact trade sanctions.
- Enact restrictions on technology transfer.
- Cancel or restrict US-funded programs.
- Reduce security assistance programs.

Figure F-4. Examples of Requested Economic Flexible Deterrent Options

2. **Description of Deterrent Actions**

a. **Deterrence is the prevention of an adversary's undesired action.** Deterrence is a state of mind brought about by the adversary's perception of three factors: the likelihood of being denied the expected benefits of his action, the likelihood of having excessive costs imposed for taking the action, and the acceptability of restraint as an alternative. These effects are the results of a synchronized and coordinated use of all instruments of national power. **FDOs are deterrent-oriented response options** that are requested and may be initiated based on evaluation of indicators of heightened regional tensions.

b. **FDOs serve two basic purposes. First,** they provide a visible and credible message to shape adversary perceptions about the costs and benefits of undesired activity. **Second,** they position US forces in a manner that facilitates implementation of OPLANs/CONPLANs or OPORDs if hostilities are unavoidable. They also facilitate an early decision by laying out a wide range of interrelated response paths that are carefully tailored to avoid the classic response of "too much, too soon, or too little, too late." They are initiated before and after unambiguous warning. **Although they are not intended to place US forces in jeopardy if deterrence fails, risk analysis should be an inherent step in determining which FDO to use and how and when that FDO should be used.** FDOs have the advantage of rapid de-escalation if the situation precipitating the FDO changes.

3. **Flexible Deterrent Option Implementation**

a. The President or SecDef directs FDO implementation, and the specific FDO or combination selected will vary with each situation. Their use will be consistent with the NSS. FDOs can be used individually, in packages, sequentially, or concurrently, but are primarily developed to be used in groups that maximize integrated results from all the diplomatic, informational, military, and economic instruments of national power. It is imperative that extensive, continuous coordination occurs with interagency and multinational partners to maximize the impact of FDOs.

b. On execution of FDOs, the commander and staff must conduct assessments to determine if the objectives of the plan need to be changed to accommodate the new conditions in the OE. It is possible deterrence prevented escalation or further aggression without returning conditions to pre-crisis state. In this case, commanders need to consult with their leadership to determine new objectives.

SECTION B. FLEXIBLE RESPONSE OPTIONS

4. General

A FRO is an operational- to strategic-level concept of operation that is easily scalable, provides military options, and facilitates rapid decision making by national leaders in response to heightened threats or attacks against the US homeland or US interests. They are usually used for response to terrorist actions or threats.

5. Description of Flexible Response Options

a. The basic purpose of FROs is to preempt and/or respond to attacks against the US and/or US interests. FROs are intended to facilitate early decision making by developing a wide range of prospective actions carefully tailored to produce desired effects, congruent with national security policy objectives. A FRO is the means by which various military capabilities are made available to the President and SecDef, with actions appropriate and adaptable to existing circumstances, in reaction to any threat or attack.

b. FROs are used to address both specific, transregional threats and nonspecific, heightened threats. FROs are operations that are first and foremost designed to preempt enemy attacks, but also provide DOD the necessary planning framework to fast-track requisite authorities and approvals necessary to address dynamic and evolving threats.

c. FROs are developed as directed by the CJCS and maintained by the CCMDs to address the entire range of possible threats. FROs should support both long-term regional and national security policy objectives. Initially, FROs are developed pre-crisis by CCMDs, based on intelligence collection and analysis and critical factors analysis, and then modified and/or refined or developed real-time. FRO content guidelines are listed in Figure F-5.

d. FROs should not be limited to current authorities or approvals; rather, planning should be based on DOD's capabilities (overt, clandestine, low visibility, and covert) to achieve objectives, independent of risk. While entirely unconstrained planning is not realistic or prudent, the intent of FROs is to provide national leaders a full range of military options to include those prohibited in the current OE. Planning must also identify expected effects (to include effects on third parties, partners, and allies), resources required, and risk associated with each option.

```
Flexible Response Option Content Guidelines

•  Identify critical enemy vulnerabilities and specific targets for each major vulnerability
•  Operation objectives
•  Desired effects
•  Essential tasks
•  Major forces and capabilities required
•  Concept of deployment
•  Concept of employment to include phasing, timing, major decision points, and
   essential interagency supporting actions
•  Concept for sustainment
•  Estimated time to achieve objectives
•  Military end state(s)
•  Additional resources or shifts essential for execution
•  Additional recommended changes in authority and approval required
•  Additional risks associated with execution and mitigation approaches
```

Figure F-5. Flexible Response Option Content Guidelines

e. FROs are divided into three broad categories. These planning categories determine the scope of FRO planning efforts:

(1) Interdict terrorist or proxy organizations to deny a subgroup, affiliate, and ally or network the capability to function with global reach, access, and effectiveness.

(2) Interdict safe haven to deny the enemy and associated networks specific geographic safe haven and/or support bases.

(3) Interdict enemy critical network capabilities to deny the enemy specific functional capabilities.

f. **FRO Characteristics**

(1) Provides military options to national leadership.

(2) Military CONOPS at the operational or strategic level.

(3) Provides a start point for iterative planning.

(4) Scalable based on situation and SecDef guidance.

(5) Focused on enemy critical vulnerabilities.

(6) Nested with national and regional strategy.

(7) Deliberate and synchronized expansion of the campaign rather than disparate actions.

(8) A combination of direct and indirect actions.

(9) Decisive action or set conditions for follow-on operations.

(10) Must include a discussion of risk and probability of escalation.

6. Flexible Response Option Implementation

a. The planning engine for FROs is the contingency planning process. In the event SecDef directs the execution of a FRO, the supported CCMD would initiate planning to determine existing options or develop new ones for SecDef and to enable acquisition of authorities and approvals necessary to conduct appropriate military operations to disrupt terrorist threats and/or respond to attacks on the US or US interests.

b. **Applications of FROs**

(1) **Disrupt** is used to address both specific, transregional threats and nonspecific, heightened threats. Disrupt options are developed to preempt enemy attacks.

(a) **Specific Threats.** Disrupt contingencies are triggered by specific warning intelligence or identified attack plans spanning more than one AOR or otherwise requiring global integration, as determined by CJCS.

(b) **Nonspecific Threats.** Disrupt is also triggered by general indications of increased terrorist threats, in the absence of actionable intelligence against a specific threat. Periodically, intelligence assessments indicate enemy strength has increased despite current operations or terrorist attack preparations have progressed to the point that national leadership is willing to consider additional operations, actions, and activities.

(2) **Response.** Respond contingencies are triggered as a result of a successful or unsuccessful attack against the US or its interests. If efforts fail to preempt, disrupt, or defeat a major attack, respond options rapidly provide flexible and scalable options to respond with global operations against the entire scope of the enemy (see Figure F-6). The following are examples of FRO scalability. Operations in each category can be executed individually, concurrently, or sequentially.

(a) **Rapid Response.** Priority of effort is to demonstrate US resolve through speed of action. Rapid responses would most likely be unilateral strikes, raids, cyberspace operations, and IO against known targets with low collateral damage.

(b) **Limited Response.** Priority of effort is to attack organizations directly attributed to the attack. The objective of this category is to maximize perceived legitimacy of US response. Limited response demonstrates restraint and is more likely to garner international cooperation. Disadvantages may include uncertain timeline due to

Flexible Response Option Scalability

Rapid Response	Limited Response	Decisive Response
Demonstrate Resolve	Target Those Directly Responsible	Defeat Violent Extremist Organization
Priority of Effort: • Speed	**Priority of Effort:** • Legitimacy via attribution	**Priority of Effort:** • Direct attack on enemy center of gravity
Advantages: • Demonstrate resolve • Least impact of current operations	**Advantages:** • Response aimed directly at those responsible • Demonstrates restraint • International cooperation more likely	**Advantages:** • Proactive vice reactive • Targets critical enemy vulnerabilities • Greater impact on enemy
Disadvantages: • Limited strategic effect • More likely lethal in nature • Probable negative international reaction • More likely unilateral action	**Disadvantages:** • Uncertain timeline • Persistent operation may require reallocation of resources • United States remains vulnerable to other extremist organization elements	**Disadvantages:** • Potential to destabilize region of focus • Perception of US overreaction • Higher risk • Unintended consequences

Figure F-6. Flexible Response Option Scalability

requirement for attribution and continued vulnerability to networks not directly associated with the current attack.

(c) **Decisive Response.** Priority of effort is to attack the enemy operational COG to achieve a long-term disruption of its operational capability. This category is proactive vice reactive and seeks greater long-term impact on or defeat of the enemy. Disadvantages may include perception of US overreaction with possible negative public opinion consequences and the potential provocation of retaliatory responses of various kinds.

APPENDIX G
COURSE OF ACTION COMPARISON

The most common technique for COA comparison is the weighted numerical comparison, which uses evaluation criteria to determine the preferred COA based upon the wargame. COAs are not compared to each other directly until each COA is considered independently against the evaluation criteria. The CCDR may direct some of these criteria, but most criteria are developed by the JPG as detailed in Chapter V, "Joint Planning Process." Below are examples of common methods.

1. Weighted Numerical Comparison Technique

a. The example below provides a numerical method for differentiating COAs. Numerical methods are often mathematically deficient and can lead to incorrect or false conclusions (particularly given the inherently subjective numerical assignments). Experienced planners avoid numerical COA comparison methodology as overly simplistic. Values reflect the relative preference of each COA within each criterion. All criteria have been weighted to reflect their relative preference to one another (Figures G-1 and G-2).

b. Recall the weight of each criterion determined in COA analysis. The staff leader responsible for a functional area scores each COA using those criteria. Multiplying the score by the weight yields the criterion's value. The staff leader then totals all values. The staff member must not portray this simplified numeric method as the result of a rigorous mathematical analysis. Comparing COAs by criterion is more accurate than comparing total values.

(1) Evaluation criteria are those selected through the process described in Chapter V, "Joint Planning Process."

(2) The evaluation criteria can be weighted. The most important criteria are rated with the highest numbers. Lesser criteria are weighted with progressively lower numbers.

(3) The highest number is best. The best criterion and the most advantageous COA ratings are those with the highest number. Values reflect the relative strengths and weaknesses of each COA.

(4) Each staff section does this separately, perhaps using different criteria on which to base the COA comparison. The staff then assembles and arrives at a consensus for the criterion and weights. The COS or JTF deputy commander should approve the staff's recommendations concerning the criteria and weights to ensure completeness and consistency throughout the staff sections.

2. Non-Weighted Numerical Comparison Technique

The same as the previous method except the criteria are not weighted. Again, the highest number is best for each of the criteria.

Example Numerical Comparison

Criteria	Weight	Course of Action					
		COA 1		COA 2		COA 3	
		Rating	Product	Rating	Product	Rating	Product
Exploits maneuver	2	3	6	2	4	1	2
Attacks COGs	3	2	6	3	9	1	3
Integrates maneuver and interdiction	2	2	4	3	6	1	2
Exploits deception	2	1	2	2	4	3	6
Provides flexibility	2	1	2	3	6	2	4
CSS (best use of transportation)	1	3	3	2	2	1	1
Total		12		15		9	
Weighted total			23		31		18

NOTE: The higher the number, the better.

- The joint force commander's intent explained that the most important criterion was "attacking the enemy's COGs." Therefore, assign a value of 3 for that criterion and lower numbers for other criteria that the staff devises (**this is the weighing criterion**).

- For attacking the enemy COGs, COA 2 was rated the best (with a number of 3). Therefore, COA 2 = 9, COA 1 = 6, and COA 3 = 3.

- After the relative COA **rating** is multiplied by the **weight** given each criterion and the product columns are added, COA 2 (with a score of 31) is rated the most appropriate according to the criteria used to evaluate it.

Legend

COA   course of action   COG   center of gravity   CSS   combat service support

Figure G-1. Example Numerical Comparison

Example #2 Course of Action Comparison Matrix Format

Evaluation Criterion	Weight	COA 1		COA 2		COA 3	
		Score	Weighted	Score	Weighted	Score	Weighted
Surprise	2	3	6	1.5	3	1.5	3
Risk	2	3	6	1	2	2	4
Flexibility	1	3	3	1.5	1.5	1.5	1.5
Retaliation	1	1.5	1.5	3	3	1.5	1.5
Damage to alliance	1	3	3	1.5	1.5	1.5	1.5
Legal basis	1	2	2	3	3	1	1
External support	1	3	3	2	2	1	1
Force protection	1	2.5	2.5	2.5	2.5	1	1
OPSEC	1	3	3	1.5	1.5	1.5	1.5
Total			30		20		16

NOTE: The higher the number, the better.

Legend

COA   course of action   OPSEC   operations security

Figure G-2. Example #2 Course of Action Comparison Matrix Format

3. Narrative or Bulletized Descriptive Comparison of Strengths and Weaknesses or Advantages and Disadvantages

Summarize comparison of all COAs by analyzing strengths and weaknesses or advantages and disadvantages for each criterion. See Figures G-3 and G-4 for examples.

Criteria for Strengths and Weaknesses Example

	Criteria 1		Criteria 2		Criteria 3	
	Strengths	Weaknesses	Strengths	Weaknesses	Strengths	Weaknesses
COA 1	• •	• •	• •	• •	• •	• •
	Strengths	Weaknesses	Strengths	Weaknesses	Strengths	Weaknesses
COA 2	• •	• •	• •	• •	• •	• •
	Strengths	Weaknesses	Strengths	Weaknesses	Strengths	Weaknesses
COA 3	• •	• •	• •	• •	• •	• •

Legend

COA   course of action

Figure G-3. Criteria for Strengths and Weaknesses Example

Descriptive Comparison Example

	Criteria 1		Criteria 2		Criteria 3	
	Advantages	Disadvantages	Advantages	Disadvantages	Advantages	Disadvantages
COA 1	• •	• •	• •	• •	• •	• •
	Advantages	Disadvantages	Advantages	Disadvantages	Advantages	Disadvantages
COA 2	• •	• •	• •	• •	• •	• •
	Advantages	Disadvantages	Advantages	Disadvantages	Advantages	Disadvantages
COA 3	• •	• •	• •	• •	• •	• •

Legend

COA   course of action

Figure G-4. Descriptive Comparison Example

4. Plus/Minus/Neutral Comparison

Base this comparison on the broad degree to which selected criteria support or are reflected in the COA. This is typically organized as a table showing (+) for a positive influence, (0) for a neutral influence, and (–) for a negative influence. Figure G-5 is an example.

5. Descriptive Comparison

This is simply a description of advantages and disadvantages of each COA. See Figure G-4.

Plus/Minus/Neutral Comparison Example

Criteria	COA 1	COA 2
Casualty estimate	+	–
Casualty evacuation routes	–	+
Suitable medical facilities	0	0
Flexibility	+	–

Legend

COA   course of action

Figure G-5. Plus/Minus/Neutral Comparison Example

Intentionally Blank

APPENDIX H
POSTURE PLANS

1. Overview

a. Posture plans are key elements of CCMD campaigns and strategies. They describe the forces, footprint, and agreements the commander needs in order to successfully execute the campaign.

b. **Posture Plans**

(1) GCCs prepare TPPs, as directed in the GEF and JSCP, which outline their posture strategy, link national and theater objectives with the means to achieve them, and identify posture requirements and initiatives to meet campaign objectives.

(2) The TPP is the primary document used to advocate for changes to posture and to support resource decisions, the posture management process, and departmental oversight responsibilities. It delineates the CCMD's posture status, with gaps, risks, and required changes substantiated by national and theater strategy, and proposes initiatives that address challenges. The status of the CCMD's compliance with GEF and JSCP posture guidance should be clearly articulated.

(3) GCCs' TPPs also address the overseas posture requirements of other DOD stakeholders in theater. Stakeholders include other geographic CCMDs, the functional CCMDs, the Military Departments, DOD agencies, and non-DOD agencies and field activities. GCCs coordinate their TPPs with these stakeholders to incorporate their requirements.

(4) Functional CCMDs prepare functional posture plans to enable their assigned missions and support the GCCs. USSOCOM prepares a Global Special Operations Forces Posture Plan, USSTRATCOM prepares a Strategic Infrastructure Master Plan, and USTRANSCOM prepares its En Route Infrastructure Master Plan. Each plan includes a strategic narrative that assesses posture gaps, associated risks, and posture initiatives recommended to address those gaps/risks. These plans are coordinated with the geographic CCMDs to facilitate consideration and incorporation of functional requirements into the TPPs.

c. **Process**

(1) GDP is managed by the OUSD(P) and the JS. CCMD posture plans and master plans outline the current posture and propose posture initiatives for a two-to-five year timeframe and beyond. Campaign planners are informed by posture subject matter experts on strategic and operational access issues. In turn, the strategy and the operational approach of the campaign plan inform the posture plan.

(2) The Global Posture Executive Council (GPEC) is DOD's senior posture governance body. The GPEC facilitates senior leader posture decision making; enables the CCMDs, Military Departments and Services, and DOD agencies to collaborate in DOD's

GDP planning; and oversees the implementation and assessment of DOD's posture plans. The JS J-5, in coordination with OUSD(P), annually provides GPEC-endorsed posture guidance to guide the development of posture plans.

2. Elements of a Posture Plan

a. **Forces.** Forces are composed of assigned, allocated, and enabling units, personnel, and assets. It includes rotational and mobility forces. They execute the mission through offensive, defensive, and stability activities.

b. **Footprint.** The footprint includes enduring locations, supporting infrastructure, and prepositioned equipment.

c. **Agreements.** Agreements provide access, basing, lawful mission execution, protection, and relationships which allow the footprint to be established and forces to execute their missions. Examples are access agreements, basic ordering agreements, transit agreements, status-of-forces agreements, and treaties.

3. Posture Terminology

a. **GDP Locations:**

(1) **Contingency Location.** A non-enduring location outside of the US that supports and sustains operations during named and unnamed contingencies or other operations as directed by appropriate authority and is categorized by mission life-cycle requirements as initial, temporary, or semi-permanent.

(2) **Enduring Location.** DOD uses an established lexicon for the types of overseas (in foreign countries or US territories overseas) locations from which it operates in its GDP framework. A location is enduring when DOD intends to maintain access and or use of that location for the foreseeable future. Enduring locations play a critical role in allowing DOD to deploy, or employ, US forces when and where necessary, but need not have a continuous force presence or permanent structure sustained or constructed through US appropriations. CCDRs nominate locations as enduring. The GPEC reviews and endorses the list and OSD validates these nominations in consultation with DOS. The following types of sites are considered enduring for USG purposes: CSL, FOS, and MOB. All three types of locations may be composed of more than one distinct site.

(a) **CSL.** An enduring GDP location characterized by the periodic presence of rotational US forces, with little or no permanent US military presence or US-owned infrastructure, used for a range of missions and capable of supporting requirements for contingencies. CSLs may feature a small permanent presence of assigned support personnel (military or contractor). CSLs typically consist of mostly HN infrastructure, and CSL real property is often not US-owned (i.e., not part of the US real property inventory). However, CSLs may require US-funded infrastructure to meet operational requirements. CSLs are a focal point for security cooperation activities and provide contingency access, logistic support, and rotational use by operational forces, and can support an increased force presence during contingencies of finite duration.

(b) **FOS.** An enduring GDP location characterized by the sustained presence of rotational US forces, with infrastructure and quality of life amenities consistent with that presence, capable of providing forward staging for operational missions and support to regional contingencies. FOSs consist of US-owned real property, and they may feature a small permanent presence of assigned support personnel (military or contractor). FOSs often support the stationing of pre-positioned force, equipment, or supplies, and they can serve as a regional hub in support of regional contingencies.

(c) **MOB.** An enduring GDP location characterized by the presence of permanently assigned US forces and robust infrastructure that typically includes C2, highly developed force protection measures, hardened facilities, and significant quality of life amenities, often including family support facilities. MOBs consist of US-owned real property and represent primary training and deployment locations for the US overseas. MOBs can support both small and large scale operations and global contingencies.

(3) **En Route Location.** A term specific to global distribution: an intermediate node outside of the CONUS distribution network that supports refueling, maintenance, crew change/rest, or transload of cargo/passengers for onward movement to final destination. These nodes can be a fixed or a temporary location. En route locations may be designated as MOBs, FOSs, or CSLs.

b. **Other Terms**

(1) **Posture Initiative.** A required change to US GDP that meets any one of the following criteria:

(a) Has policy significance.

(b) Comprises a significant change to an element of defense posture (forces, footprint, or agreements).

(c) Comprises a significant change in capability.

(d) Has significant impact on resourcing.

(2) **Policy Significance.** A change to US GDP that has policy significance is one that would, among other things, directly and significantly affect foreign or defense relations between the US and another government; would require approval, negotiation, or signature at the OSD or diplomatic level of an international agreement or other document relating to US forces' presence in a foreign country; would generate a major increase or decrease in US operational or sustainment capability at one or more locations; or would result in the designation of a new enduring location or re-designation of an existing enduring location (as an Enduring Locations Master List change nomination). The phrase "policy significance" should be interpreted broadly in this context. A determination by the USD(P) that a particular change to US GDP has policy significance is conclusive.

(3) **Presence.** Presence is the physical location of forces at a location. Presence can be permanent or temporary.

4. Format for Posture Plans

SAMPLE POSTURE PLAN

a. Executive Summary

b. Strategic Narrative. This should mirror the theater or functional strategy prepared by the CCDR.

c. Forces

    (1) Approach to Force Posture.

    (2) Gaps and Risk and Implications.

    (3) Recommended Force Changes.

d. Footprint

    (1) Approach to Footprint. Identify the specific posture approach for footprint (e.g., bases, locations) in the region. Discuss how the current footprint supports Theater and National objectives.

    (2) Gaps and Risk.

    (3) Footprint Changes Required.

    (4) Contingency Locations.

e. Agreements

    (1) Approach to Agreements.

    (2) Status of Current Agreements.

    (3) Gaps and Risk.

    (4) Agreements in Development.

        (a) Status of Agreements in Development.

        (b) Priority of need of Agreements in Development.

    (5) Host Nation Agreements.

    (6) Basic ordering agreements (e.g., current contracts or task orders).

f. Unresolved Policy Issues

g. Appendices. The TPP should include:

    (1) Theater Maps.

(2) Charts of Footprint by Location.

(3) Project Summary Sheets.

(4) Enduring Location Changes.

For detailed information and guidance on GDP and posture plans, see GPEC Posture Planning Guidance and supplemental guidance to the JSCP, CJCSI 3110.01, Joint Strategic Capabilities Plan.

Intentionally Blank

APPENDIX J
THEATER DISTRIBUTION PLANS

1. Overview

a. TDPs describe the distribution network within each of the geographic CCMDs' AOR (outside the continental US) as directed by the GEF and JSCP. They describe the distribution pipeline from the point of need to the point of employment.

b. USTRANSCOM, as the global distribution synchronizer, provides a TDP template in the *Campaign Plan for Global Distribution* and will advise and assist the GCCs with the development and improvement of their TDPs on a biennial cycle.

c. TDPs provide detailed theater mobility and distribution analysis to assist in planning current and future operations, inform the TCP and other plans, and aids theater distribution decision making.

d. **Distribution Plans**

(1) Geographic CCMDs prepare TDPs, as directed in the GEF, JSCP, and JSCP Logistics Supplement. TDPs ensure sufficient distribution capacity throughout the theater and synchronization of distribution planning throughout the global distribution network. This synchronization enables a GCC's theater distribution to support the development of TCPs and OPLANs.

(2) GCC TDPs should also address the overseas distribution requirements of other DOD stakeholders in theater. Stakeholders include other geographic CCMDs, the functional CCMDs, the Military Departments, DOD agencies, and non-DOD agencies. GCCs coordinate their TDPs with these stakeholders to incorporate their requirements.

e. **Process**

(1) The Under Secretary of Defense for Acquisition, Technology, and Logistics is responsible for establishing policies for logistics, maintenance, and sustainment for all elements of the DOD. The USDP provides oversight of the Campaign Plan for Global Distribution. The *Campaign Plan for Global Distribution* and CCMD distribution plans outline the current and proposed distribution issues for near, mid, and far term timeframes. Campaign planners are informed by posture plans which provide the foundation for distribution subject matter experts to layout the distribution network through TDPs into their AOR.

(2) The *Campaign Plan for Global Distribution* annual synchronization seminar is the DOD's senior distribution governance forum. The synchronization seminar facilitates senior leader distribution issue resolution decision-making process with a collaborative effort among the CCMDs, the Services, and DOD agencies for implementation and assessment of DOD's distribution plans.

(3) The Commander, USTRANSCOM, in coordination with OUSD(P), annually provides a global distribution assessment of issues that affect the global distribution network.

2. Elements of a Theater Distribution Process

a. The TDP contains detailed information on the theater distribution capabilities and their interface with the global distribution network for a GCC's AOR. It reflects the theater's physical means, processes, people, and systems required for the receipt, storage, staging, and movement of forces and materiel from points of origin to points of employment. The TDP provides theater intelligence, as well as transportation and capacity specific information on ports, airfield, ground and sea LOCs, and distribution infrastructure within the AOR.

b. The template in the *Campaign Plan for Global Distribution* provides more detail on formatting guidance to the GCC's TDP writing teams.

3. Format for Distribution Plans

The TDP template in the *Campaign Plan for Global Distribution* is patterned after the CJCSM 3130.03, *Adaptive Planning and Execution (APEX) Planning Formats and Guidance.* USTRANSCOM is responsible for the strategic movement from point of embarkation to point of debarkation. The distribution interface with the GCCs and Services at the point of need to the point of employment is crucial for the warfighter to maintain momentum while consolidating and reorganizing forces/equipment. The TDP provides a tactical view of the distribution network (physical, communication, information, and financial) including theater specific deployment considerations. The sections are listed as follows:

a. BPLAN.

b. Task organization.

c. Intelligence.

d. Operations.

e. Logistics.

f. Meteorological and oceanographic operations.

g. HNS.

h. Medical services.

i. Reports.

j. Interagency.

k. OCS.

APPENDIX K
RED TEAMS

1. Background

The use of red teams contributes to reduce risk, avoid surprise, see opportunities, increase operational flexibility, broaden analysis, and enhance decision making. Red teams help organizations adapt to change and improve military planning and intelligence analysis by stimulating critical and creative thought. The JS has recommended the routine employment of red teams.

2. The Red Team Overview

a. All organizations, staff agencies, and work groups that study issues, draw conclusions, make plans, develop concepts, produce intelligence, create scenarios, conduct experiments, simulate adversaries, make recommendations or decide issues, can benefit from red teams. Red teams can help any staff element frame problems, challenge assumptions, counter institutional biases, stimulate critical and creative thought, and support decision making in every organization.

b. Joint forces should use red teams to enhance and complement regular processes and help ensure that all aspects of key problems are understood and the fullest range of potential options are considered.

c. Implicit tasks include countering the influence of institutional and individual bias and error; providing insight into the mindsets, perspectives, and cultural traits of adversaries and other relevant actors; and helping explore unintended consequences, follow-on effects, and unseen opportunities and threats. Red teams reduce risk by helping organizations anticipate, understand, prepare, and adapt to change.

KEY TERM

Red Team: An organizational element comprised of trained and educated members that provide an independent capability to fully explore alternatives in plans and operations in the context of the operational environment and from the perspective of adversaries, partners, and others.

---

d. The red team is a specially-trained decision-support staff asset that can be employed throughout the joint force. The red team can complement all staff problem-solving and analytical efforts by serving as a "devil's advocate" and generalized contrarian, but is normally focused on supporting plans, operations, and intelligence.

(1) **Plans and Operations.** The red team supports planning and mission execution. This includes helping identify vulnerabilities, opportunities, and faulty or unstated assumptions; helping ensure all aspects of the OE are fully understood; and critically reviewing strategies, operational concepts, estimates, plans, and orders.

(2) **Intelligence.** The red team complements intelligence efforts by offering independent, alternative assessments and differing interpretations of information. This includes critical reviews of intelligence products; considering problem sets from alternative perspectives; and helping contribute informed speculation when reliable information is lacking.

3. Red Team Support to Joint Planning

a. **Support to joint planning is a core red team function.** Organizations that plan and execute operations may employ red teams to help them think critically and creatively and see planning issues and proposed COAs from alternative perspectives. The red team may also help the staff avoid common sources of error.

b. The red team could be used to support virtually all aspects of JPP. However, the team's capacity will seldom match the scale of requirements. Accordingly, this guidance focuses on those functions where red teams can have the greatest impact on planning. **Participation in JPP, Step 2 (Mission Analysis), is normally the most effective use of the team in joint planning.**

c. Red team support to joint planning is usually provided via active participation in planning groups and the production of tailored papers and briefings that support the planning effort. When addressing key issues that may have wide-ranging effects on planning, it may be prudent to circulate comprehensive stand-alone red team products for the staff's review. While the red team may suggest alternatives for consideration, these inputs should be weighed and either incorporated or set aside as appropriate before planning products are finalized. Critical red team observations may be, at the planners' discretion, developed into branch plans.

d. During multinational planning efforts, red teams should ensure foreign staff officers understand the red team's role as a "devil's advocate," so they understand the purpose of the team's "contrarian perspective." It should also be noted that some foreign services may have their own style of red teaming, and they might be able to make valuable contributions to the overall planning efforts once their red teaming efforts are integrated with those of their US counterparts.

e. The strategic estimate is used to develop campaign plans. It encompasses all aspects of the commander's OE and is the basis for the development of the CCMD's theater or functional strategy. It provides the commander's perspective of the strategic and operational levels of the OE, desired changes required to meet specified regional or functional objectives, and the commander's visualization of how those objectives might be achieved. It addresses a number of core issues which can benefit from red team scrutiny, such as an analysis of all states, groups, or organizations in the OE; a review of relevant geopolitical, economic, and cultural factors; an assessment of strategic and operational challenges facing the CCMD; an analysis of known or anticipated opportunities; and an assessment of risks inherent in the OE. Accordingly, the draft strategic estimate should be carefully reviewed by the red team.

4. Joint Planning Activities, Functions, and Products

a. Planning is largely by intelligence assessments of current or projected situations and threats. It is generally not practical or advisable for the red team to offer independent, alternative assessments for each intelligence estimate, but within its capability, the red team should judiciously review key assessments and estimates, and, when appropriate, suggest alternatives assessments to alert the staff to previously unseen threats and opportunities that may require new planning initiatives.

b. The red team supports all planning functions through active participation in planning teams and critical reviews of draft planning materials. The red team should participate in the early stages of planning functions to ensure the staff has sufficient time to consider the red team's inputs before key decisions are made. The red team should not produce duplicative or competing planning materials, but should instead seek to incorporate its inputs into the planning team's final products. In some cases, however, it may be useful to circulate comprehensive stand-alone "think pieces" to help the staff consider specific issues, but these should never be cast as criticisms of the planning team's products.

c. Support to IPRs. Planning is facilitated by periodic IPRs that provide up- and down-channel feedback, shaping and refining as the plan is developed. Ideally, prior to an IPR, the red team should review their organizations' draft IPR briefings and papers and offer suggestions as appropriate. The red team's most critical contributions to any new plan will usually come during mission analysis, although preparations for later IPRs may actually involve more of the team's time and resources.

d. The red team should be fully integrated into the planning process and assist in the initial development and revision of JPP products. When the red team is unable to support all aspects of a specific planning effort, the commander or J-5 should establish priorities for red team support. In most cases, the red team will have the greatest impact on planning during JPP Step 2 (Mission Analysis), and Step 4 (COA Analysis and Wargaming).

e. Potential red team roles in planning are outlined below:

(1) **Step 1 (Planning Initiation).** Commanders typically provide initial planning guidance to planning teams. The red team typically participates in the planning team's review of that guidance and recommending refinements back to the commander.

(2) **Step 2 (Mission Analysis)**

(a) One key input to mission analysis is JIPOE. If the red team has not participated in the JIPOE process, then it should conduct an independent, alternative assessment of the adversary's COG, critical capabilities, and critical vulnerabilities. The red team should then offer its alternative assessments for consideration by both the intelligence staff and the planners.

(b) One primary red team task during mission analysis is to help the planners frame the problem, define desired end states, and assess known facts and key assumptions.

The red team should challenge weak assumptions or suspect facts, and, as the situation evolves, consider whether the assumptions remain valid.

(c) If possible, the red team should help determine operational limitations, termination criteria, military end state, military objectives, and mission success criteria, providing alternative perspectives and exploring how political will and cultural viewpoints might constrain operations and limit options. In addition, the team may also participate in developing specified, implied, and essential tasks, conducting risk analysis, writing the CCIRs, and drafting the mission statement. If possible, the team should participate in drafting the mission analysis brief and the commander's refined planning guidance.

(3) **Step 3 (COA Development).** The red team can often make useful contributions to COA development by helping the planners expand the range of COAs under consideration.

(4) **Step 4 (COA Analysis and Wargaming)**

(a) During COA analysis, the red team should advise planners of the potential cultural implications associated with each COA, and should help explore the potential unintended consequences and likely second and third-order effects associated with each COA.

(b) During COA wargaming, the red team should help both the simulated friendly force and the opposition/adversary force (red cell) consider the widest range of options during their respective moves. The red team should also advise both sides regarding how their moves might be perceived by relevant actors or impacted by wildcard events.

(5) **Step 5 (COA Comparison).** COA comparison is often seen as an objective measurement of the relative merits of the COAs developed and analyzed in earlier steps. The red team should participate in the development of the COA comparison criteria and highlight those areas in which subjective and cultural issues might outweigh more tangible, more easily quantified factors.

(6) **Step 6 (COA Approval).** Planners should consider including a summary of wildcards, unintended consequences, and second and third order effects in the COA decision briefing to the commander.

(7) **Step 7 (Plan or Order Development)**

(a) If the decision is to develop an order, and the red team has been actively participating to this point, then order development may continue without additional red team input. Future revisions, however, should be supported by the red team. If the red team has not been involved in planning prior to this point, then the team should review assumptions and evaluate the potential impact of cultural factors. In addition, the team should explore likely unintended consequences, second and third order effects, and wildcards. Revising the order at this point can be extremely disruptive. Accordingly, if the red team's review suggests serious shortcomings, the senior planner should be advised.

(b) If the decision is to develop a plan, then continuing red team support will be required. The team should participate in developing appendixes, making certain themes, messages, and media are compatible with the mindsets of relevant actors and that potential unintended consequences are explored. The team should also participate in developing the essential elements of information appendix and assessment annex, using its understanding of the OE to help ensure the relevance of the measurements.

(c) The OE and situation may evolve as the plan is drafted, and the red team should remain sensitive to developing threats and opportunities, wildcards, and other issues. In addition, plans typically address more issues, and in more detail, than were addressed during the working groups, and these can usually benefit from red team support.

(d) The red team should review key sections of the plan and offer recommendations while those sections are still in the draft stage. If the red team has not been involved in the planning effort before the decision brief, a red team review of these draft sections is critical.

(e) During plan development, it may be useful for the red team to circulate a document that consolidates its observations and concerns. This document can provide complete, fully reasoned descriptions of issues the red team might have previously raised in closed work groups, or it may propose new issues that planners should consider. Issues could include potential wildcards and low probability/high impact events, likely unintended consequences and second and third order effects, unseen threats and opportunities, and so on. In some cases, it may be prudent to include this document in the final plan as a reference for mission execution or to support future plan revisions, but in all cases, it must be understood that the primary purpose of the document is to support the development of the plan rather than serve as an after-the-fact critique.

(f) Completed plans are frequently refined or adjusted over time, and refinement continues even after execution. During refinement and adaptation, the red team helps assess the situation, develop new guidance, and support continued planning efforts.

5. Joint Planning During Execution

a. Red team support during mission rehearsal generally parallels that of wargaming. During rehearsal, the primary objective is to test the plan's CONOPS and COAs. As the plan is rehearsed, the red team should focus on helping the staff uncover previously unseen weaknesses, opportunities, and unintended effects. During rehearsal, the red team should be attuned to potential alternative COAs and assessments, which it may propose after rehearsal, when the staff may be actively seeking improvements or alternatives to the plan.

b. A crisis action team (CAT) is often stood up during the initial stages of a crisis. While not part of its critical analysis function, the red team may support the CAT by providing expertise in alternative interpretations of dynamic, uncertain situations, by helping frame problems, and by broadening the search for potential responses. A CAT normally uses streamlined decision-making procedures, and the primary red team mode of support will often consist of active participation in work groups rather than formal written products.

c. Planning continues throughout execution in three venues, each focused on distinct but overlapping timeframes: future plans, future operations, and current operations. The red team plays distinct roles in each of these, but should normally concentrate its efforts in future plans.

(1) Future plans addresses the next phase of operations or sequels to the current operation. It is usually conducted by the J-5, by a JPG, or, in some commands, by a long-range planning element. Future planners look for opportunities or challenges that might require a revision to the current mission or a different operational approach. Red team support to future plans will generally follow that provided during JPP Step 2 (Mission Analysis), but in an abbreviated form.

(2) Future operations addresses branches to current, on-going operations. It is normally addressed by the J-3, or, in some commands, an operations planning element. Red team support to future operations will often resemble that of future plans, but with a more truncated time horizon and more streamlined processes.

(3) Current operations addresses immediate or very near-term issues associated with ongoing operations. Current operations are usually addressed by the organization's joint operations center. Due to the compressed decision cycle, opportunities for the red team to influence the staff's thinking may be limited to providing alternative assessments of selected aspects of the on-going situation.

d. In some commands, a number of working groups are used to manage the flow of information to decision makers and to coordinate recurring decisions within the headquarters' battle rhythm. These working groups are often referred to as boards, cells, centers, and working groups. The red team should support the following groups (or their equivalents), if formed:

(1) Long-range planning group.

(2) Operations planning element.

(3) Commander's communication synchronization working group.

(4) CAB.

e. Assessment entails two distinct tasks: monitoring the situation and the progress of the operations, and evaluating operations against established MOEs and MOP to determine progress relative to established objectives. Dynamic interactions between friendly forces, adaptable adversaries, and populations can complicate assessment. Commanders must be attuned to changes in the OE, including the political variables in the OE and surrounding areas.

f. During assessment, the red team should analyze the situation from the perspective of the adversary and other stakeholders. The most important measure of success may be how the adversary assesses his own situation, rather than whether friendly forces are maximizing MOP and effectiveness scores. Operation assessment especially during combat operations, should be weighed against the enemy's perspective of his own condition, his own objectives, and his own unique mindset and world view. Even if all objective measurements and assessments portray the enemy as defeated, he may not believe he is beaten. For example, an enemy that has suffered extreme attrition, but can still conduct sporadic offensive operations, may see himself as heroic and undefeated, even when objective measures suggest otherwise. Overall, the red team should have access to the same information as the assessment elements, and whenever the red team's assessment of the adversary's mindset portrays a significantly different picture than that implied by assessment analyses, the red team input should be presented as a supplement to the assessment analyses.

g. As assessments and observations are translated into lessons learned, the red team's external vantage point can be invaluable. The team's relative independence will often help it see issues and potential solutions that might not be apparent to those closer to the problem. The team will also be less inhibited in highlighting issues and proposing corrective measures than staff elements that might bear some responsibility for the problem or that might be obligated to implement solutions.

6. Red Team Support to Intelligence Planning

a. The intelligence component of APEX is the IP process, and it is conducted by the organizations within DOD component of the intelligence community. IP procedures are fully integrated and synchronized joint planning. The IP process is a methodology to coordinate and integrate available defense intelligence capabilities to meet CCDR intelligence requirements. It ensures prioritized intelligence support is aligned with CCDR objectives for each phase of an operation. The Defense Intelligence Enterprise develops products (e.g., DTA, theater intelligence assessment, and national intelligence support plan) that are used by the joint force J-2 to provide the JFC and staff with situational understanding of the OE. Products developed by the CCMD J-2 during IP include intelligence estimates and the annex B (Intelligence).

b. These intelligence products provide substantial support to senior leader decision making throughout planning and execution, should be free from analytical error and organizational bias, and make certain all reasonable alternative interpretations have been considered. As such, red teams should be utilized in drafting these products.

Intentionally Blank

APPENDIX L
REFERENCES

The development of JP 5-0 is based on the following primary references:

1. **General**

    a. Title 10, USC.

    b. Title 22, USC.

    c. UCP.

    d. NSS.

    e. PPD-1, *Organization of the National Security Council System.*

    f. USAID's *3D Planning Guide: Diplomacy, Development, Defense.*

2. **Department of Defense Publications**

    a. *National Defense Strategy.*

    b. *Quadrennial Defense Review.*

    c. Secretary of Defense Memorandum, *Implementation of the Adaptive Planning Roadmap II,* March 2008.

    d. *Strategic Planning Guidance.*

    e. *GFMIG FY 2016-2017.*

    f. *2015-2017 Guidance for Employment of the Force (GEF).*

    g. Department of Defense Directive (DODD) 3000.06, *Combat Support Agencies (CSAs).*

    h. DODD 3000.07, *Irregular Warfare (IW).*

    i. DODD 3000.10, *Contingency Basing Outside the United States.*

    j. DODD 4500.09E, *Transportation and Traffic Management.*

    k. DODD 5100.01, *Functions of the Department of Defense and Its Major Components.*

    l. DODD 5100.03, *Support of the Headquarters of Combatant and Subordinate Unified Commands.*

m. DODD 5205.14, *Department of Defense Counter Threat Finance Policy.*

n. Department of Defense Instruction (DODI) 1100.22, *Policy and Procedures for Determining Workforce Mix.*

o. DODI 3020.41, *Operational Contract Support (OCS).*

p. DODI 5000.68, *Security Force Assistance (SFA).*

q. *Department of Defense Strategy for Countering Weapons of Mass Destruction.*

r. *DOD Dictionary of Military and Associated Terms*

3. Department of State Publication

Briefing, Office of the Coordinator for Reconstruction and Stabilization, United States Department of State, *Post Conflict Reconstruction: Essential Tasks.*

4. Chairman of the Joint Chiefs of Staff Publications

a. CJCSI 2300.01D, *International Agreements.*

b. CJCSI 2420.01D, *(U) United States Freedom of Navigation Program.*

c. CJCSI 3100.01C, *Joint Strategic Planning System.*

d. CJCSI 3110.01J, *(U) 2015 Joint Strategic Capabilities Plan (JSCP).*

e. CJCSI 3110.02H, *(U) Intelligence Planning Objectives, Guidance, and Tasks.*

f. CJCSI 3141.01E, *Management and Review of Joint Strategic Capabilities Plan (JSCP)-Tasked Plans.*

g. CJCSI 3150.25F, *Joint Lessons Learned Program.*

h. CJCSI 3210.06A, *Irregular Warfare.*

i. CJCSI 3401.01E, *Joint Combat Capability Assessment.*

j. CJCSI 3401.02B, *Force Readiness Reporting.*

k. CJCSI 3500.01H, *Joint Training Policy for the Armed Forces of the United States.*

l. CJCSI 3500.02B, *Universal Joint Task List Program.*

m. CJCSI 5120.02D, *Joint Doctrine Development System.*

n. CJCSI 5714.01D, *Policy for the Release of Joint Information.*

o. CJCSI 5715.01C, *Joint Staff Participation in Interagency Affairs.*

p. CJCSI 8501.01B, *Chairman of the Joint Chiefs of Staff, Combatant Commanders, Chief, National Guard Bureau, and Joint Staff Participation in the Planning, Programming, Budgeting, and Execution Process.*

q. CJCSM 3105.01, *Joint Risk Analysis.*

r. CJCSM 3122.02D, *Joint Operation Planning and Execution System (JOPES), Volume III, Time-Phased Force and Deployment Data Development and Deployment Execution.*

s. CJCSM 3130.01A, *Campaign Planning Procedures and Responsibilities.*

t. CJCSM 3130.03, *Adaptive Planning and Execution (APEX) Planning Formats and Guidance.*

u. CJCSM 3130.06A, *(U) Global Force Management Allocation Policies and Procedures.*

v. CJCSM 3150.01C, *Joint Reporting Structure General Instructions.*

w. CJCSM 3314.01A, *Intelligence Planning.*

x. CJCSM 3500.03E, *Joint Training Manual for the Armed Forces of the United States.*

y. *(U) National Military Strategy.*

5. Joint Publications

a. JP 1, *Doctrine for the Armed Forces of the United States.*

b. JP 1-0, *Joint Personnel Support.*

c. JP 2-0, *Joint Intelligence.*

d. JP 2-01.3, *Joint Intelligence Preparation of the Operational Environment.*

e. JP 3-0, *Joint Operations.*

f. JP 3-05, *Special Operations.*

g. JP 3-07, *Stability.*

h. JP 3-08, *Interorganizational Cooperation.*

i. JP 3-09, *Joint Fire Support.*

j. JP 3-11, *Operations in Chemical, Biological, Radiological, and Nuclear Environments.*

k. JP 3-12, *Cyberspace Operations.*

l. JP 3-13, *Information Operations.*

m. JP 3-14, *Space Operations.*

n. JP 3-16, *Multinational Operations.*

o. JP 3-20, *Security Cooperation.*

p. JP 3-22, *Foreign Internal Defense.*

q. JP 3-24, *Counterinsurgency.*

r. JP 3-30, *Command and Control of Joint Air Operations.*

s. JP 3-33, *Joint Task Force Headquarters.*

t. JP 3-34, *Joint Engineer Operations.*

u. JP 3-35, *Deployment and Redeployment Operations.*

v. JP 3-40, *Countering Weapons of Mass Destruction.*

w. JP 3-41, *Chemical, Biological, Radiological, and Nuclear Response.*

x. JP 3-60, *Joint Targeting.*

y. JP 3-61, *Public Affairs.*

z. JP 4-0, *Joint Logistics.*

aa. JP 4-01, *The Defense Transportation System.*

bb. JP 4-02, *Joint Health Services.*

cc. JP 4-05, *Joint Mobilization Planning.*

dd. JP 4-06, *Mortuary Affairs.*

ee. JP 4-10, *Operational Contract Support.*

ff. JP 6-0, *Joint Communications System.*

gg. Joint Doctrine Note 2-13, *Commander's Communication Synchronization.*

6. Service Publications

a. Air Force Doctrine Volume III, *Command.*

b. Air Force Doctrine Annex (AFDA) 3-0, *Operations and Planning.*

c. AFDA 3-05, *Special Operations.*

d. AFDA 3-12, *Cyberspace Operations.*

e. AFDA 3-13, *Information Operations.*

f. AFDA 3-14, *Space Operations.*

g. AFDA 3-17, *Air Mobility Operations.*

h. AFDA 3-30, *Command and Control.*

i. AFDA 3-70, *Strategic Attack.*

j. Air Force Doctrine Document 3-14.1, *Counterspace Operations.*

k. Marine Corps Doctrinal Publication (MCDP) 1-2, *Campaigning.*

l. MCDP 5, *Planning.*

m. Army Doctrine Publication 3-0, *Unified Land Operations.*

n. ADRP 3-07, *Stability.*

o. ADRP 5-0, *The Operations Process.*

p. FM 3-14, *Army Space Operations.*

q. FM 3-24, *Insurgencies and Countering Insurgencies.*

r. FM 6-0, *Commander and Staff Organization and Operations.*

s. Naval Doctrine Publication 1, *Naval Warfare.*

t. Navy Warfare Publication 3-62M, *Seabasing.*

u. *Army Mobilization and Operations Planning and Execution System.*

v. *Navy Capabilities and Mobilization Plan.*

w. *Marine Corps Capabilities Plan and Marine Corps Mobilization, Activation, Integration, Deactivation Plan.*

x. *Air Force War and Mobilization Plan.*

7. Multiservice Publication

Army Techniques Publication 5-0.3/Marine Corps Reference Publication 5-1C/Navy Tactics, Techniques, and Procedures 5-01.3/Air Force Tactics, Techniques, and Procedures 3-2.87, *Multi-Service Tactics Techniques and Procedures for Operation Assessment.*

8. Other Publications

a. *North Atlantic Treaty Organization Assessment Handbook.*

b. *Memorandum of Agreement between the Department of Defense and Department of Homeland Security on the Use of US Coast Guard Capabilities and Resources in Support of the National Military Strategy, 23 May 2008.*

APPENDIX M
ADMINISTRATIVE INSTRUCTIONS

1. User Comments

Users in the field are highly encouraged to submit comments on this publication using the Joint Doctrine Feedback Form located at: https://jdeis.js.mil/jdeis/jel/jp_feedback_form.pdf and e-mail it to: js.pentagon.j7.mbx.jedd-support@mail.mil. These comments should address content (accuracy, usefulness, consistency, and organization), writing, and appearance.

2. Authorship

The lead agent and JS doctrine sponsor for this publication is the Director for Strategic Plans and Policy (J-5).

3. Supersession

This publication supersedes JP 5-0, *Joint Operation Planning*, 11 August 2011.

4. Change Recommendations

a. To provide recommendations for urgent and/or routine changes to this publication, please complete the Joint Doctrine Feedback Form located at: https://jdeis.js.mil/jdeis/jel/jp_feedback_form.pdf and e-mail it to: js.pentagon.j7.mbx.jedd-support@mail.mil.

b. When a Joint Staff directorate submits a proposal to the CJCS that would change source document information reflected in this publication, that directorate will include a proposed change to this publication as an enclosure to its proposal. The Services and other organizations are requested to notify the Joint Staff J-7 when changes to source documents reflected in this publication are initiated.

5. Lessons Learned

The Joint Lessons Learned Program (JLLP) primary objective is to enhance joint force readiness and effectiveness by contributing to improvements in doctrine, organization, training, materiel, leadership and education, personnel, facilities, and policy. The Joint Lessons Learned Information System (JLLIS) is the DOD system of record for lessons learned and facilitates the collection, tracking, management, sharing, collaborative resolution, and dissemination of lessons learned to improve the development and readiness of the joint force. The JLLP integrates with joint doctrine through the joint doctrine development process by providing lessons and lessons learned derived from operations, events, and exercises. As these inputs are incorporated into joint doctrine, they become institutionalized for future use, a major goal of the JLLP. Lessons and lessons learned are routinely sought and incorporated into draft JPs throughout formal staffing of the development process. The JLLIS Website can be found at https://www.jllis.mil (NIPRNET) or http://www.jllis.smil.mil (SIPRNET).

6. Distribution of Publications

Local reproduction is authorized, and access to unclassified publications is unrestricted. However, access to and reproduction authorization for classified JPs must be IAW DOD Manual 5200.01, Volume 1, *DOD Information Security Program: Overview, Classification, and Declassification,* and DOD Manual 5200.01, Volume 3, *DOD Information Security Program: Protection of Classified Information.*

7. Distribution of Electronic Publications

a. Joint Staff J-7 will not print copies of JPs for distribution. Electronic versions are available on JDEIS Joint Electronic Library Plus (JEL+) at https://jdeis.js.mil/jdeis/index.jsp (NIPRNET) and http://jdeis.js.smil.mil/jdeis/index.jsp (SIPRNET), and on the JEL at http://www.dtic.mil/doctrine (NIPRNET).

b. Only approved JPs are releasable outside the combatant commands, Services, and Joint Staff. Defense attachés may request classified JPs by sending written requests to Defense Intelligence Agency (DIA)/IE-3, 200 MacDill Blvd., Joint Base Anacostia-Bolling, Washington, DC 20340-5100.

c. JEL CD-ROM. Upon request of a joint doctrine development community member, the Joint Staff J-7 will produce and deliver one CD-ROM with current JPs. This JEL CD-ROM will be updated not less than semi-annually and when received can be locally reproduced for use within the combatant commands, Services, and combat support agencies.

GLOSSARY
PART I—ABBREVIATIONS, ACRONYMS, AND INITIALISMS

ADRP	Army doctrine reference publication
AFDA	Air Force doctrine annex
AJA	annual joint assessment
AJP	Allied joint publication
ALERTORD	alert order
AOR	area of responsibility
APEX	Adaptive Planning and Execution
BPLAN	base plan
C2	command and control
CAB	commander's assessment board
CAT	crisis action team
CCDR	combatant commander
CCIR	commander's critical information requirement
CCMD	combatant command
CJCS	Chairman of the Joint Chiefs of Staff
CJCSI	Chairman of the Joint Chiefs of Staff instruction
CJCSM	Chairman of the Joint Chiefs of Staff manual
COA	course of action
COG	center of gravity
CONOPS	concept of operations
CONPLAN	concept plan
CONUS	continental United States
COS	chief of staff
CSA	combat support agency
CSL	cooperative security location
DCP	data collection plan
DEPORD	deployment order
DOD	Department of Defense
DODD	Department of Defense directive
DODI	Department of Defense instruction
DOS	Department of State
DSM	decision support matrix
DSR	defense strategy review
DST	decision support template
DTA	dynamic threat assessment
EXORD	execute order
FCC	functional combatant commander
FCP	functional campaign plan

FDO	flexible deterrent option
FFIR	friendly force information requirement
FM	field manual (Army)
FOS	forward operating site
FP	force provider
FRAGORD	fragmentary order
FRO	flexible response option
FYDP	Future Years Defense Program
GCC	geographic combatant commander
GDP	global defense posture
GEF	Guidance for Employment of the Force
GFM	global force management
GFMAP	Global Force Management Allocation Plan
GFMIG	Global Force Management Implementation Guidance
GPEC	Global Posture Executive Council
HN	host nation
HNS	host-nation support
HSC	Homeland Security Council
ID	identification
IP	intelligence planning
IPR	in-progress review
IRC	information-related capability
ISR	intelligence, surveillance, and reconnaissance
J-2	intelligence directorate of a joint staff
J-3	operations directorate of a joint staff
J-5	plans directorate of a joint staff
J-8	force structure, resource, and assessment directorate of a joint staff
JCCA	joint combat capability assessment
JCS	Joint Chiefs of Staff
JFC	joint force commander
JFLCC	joint force land component commander
JFM	joint functional manager
JFP	joint force provider
JIA	joint individual augmentation
JIOC	joint intelligence operations center
JIPOE	joint intelligence preparation of the operational environment
JOA	joint operations area
JOPES	Joint Operation Planning and Execution System
JP	joint publication
JPEC	joint planning and execution community

JPG	joint planning group
JPP	joint planning process
JRSOI	joint reception, staging, onward movement, and integration
JS	Joint Staff
JSCP	Joint Strategic Campaign Plan
JSPS	Joint Strategic Planning System
JTF	joint task force
KLE	key leader engagement
LOC	line of communications
LOE	line of effort
LOO	line of operation
LSA	logistics supportability analysis
MCDP	Marine Corps doctrinal publication
MNF	multinational force
MOB	main operating base
MOE	measure of effectiveness
MOP	measure of performance
NAI	named area of interest
NATO	North Atlantic Treaty Organization
NGO	nongovernmental organization
NMS	national military strategy
NSC	National Security Council
NSS	national security strategy
OA	operational area
OCS	operational contract support
OE	operational environment
OPLAN	operation plan
OPORD	operation order
OPT	operational planning team
OSD	Office of the Secretary of Defense
OUSD(P)	Office of the Under Secretary of Defense for Policy
PIR	priority intelligence requirement
PLANORD	planning order
PMESII	political, military, economic, social, information, and infrastructure
PPD	Presidential policy directive
PTDO	prepare to deploy order
RATE	refine, adapt, terminate, execute
RFF	request for forces

ROE	rules of engagement
RUF	rules for the use of force
SAWG	strategic assessment working group
SecDef	Secretary of Defense
SGS	strategic guidance statement
SOP	standard operating procedure
TCP	theater campaign plan
TDP	theater distribution plan
TLA	theater logistics analysis
TLO	theater logistics overview
TMM	transregional, multi-domain, and multi-functional
TPFDD	time-phased force and deployment data
TPFDL	time-phased force and deployment list
TPP	theater posture plan
UCP	Unified Command Plan
UN	United Nations
USAID	United States Agency for International Development
USC	United States Code
USCG	United States Coast Guard
USD(P)	Under Secretary of Defense for Policy
USG	United States Government
USSOCOM	United States Special Operations Command
USSTRATCOM	United States Strategic Command
USTRANSCOM	United States Transportation Command
WARNORD	warning order

PART II—TERMS AND DEFINITIONS

acceptability. The plan review criterion for assessing whether the contemplated course of action is proportional, worth the cost, consistent with the law of war, and is militarily and politically supportable. (Approved for incorporation into the DOD Dictionary.)

Adaptive Planning and Execution. A Department of Defense enterprise of joint policies, processes, procedures, and reporting structures, supported by communications and information technology, that is used by the joint planning and execution community to monitor, plan, and execute mobilization, deployment, employment, sustainment, redeployment, and demobilization activities associated with joint operations. Also called **APEX.** (Approved for replacement of "Adaptive Planning and Execution system" and its definition in the DOD Dictionary.)

adequacy. The plan review criterion for assessing whether the scope and concept of planned operations can accomplish the assigned mission and comply with the planning guidance provided. (Approved for incorporation into the DOD Dictionary.)

alert order. 1. A planning directive normally associated with a crisis, issued by the Chairman of the Joint Chiefs of Staff, on behalf of the President or Secretary of Defense, that provides essential planning guidance and directs the development, adaptation, or refinement of a plan/order after the directing authority approves a military course of action. 2. A planning directive that provides essential planning guidance, directs the initiation of planning after the directing authority approves a military course of action, but does not authorize execution. Also called **ALERTORD.** (Approved for incorporation into the DOD Dictionary.)

allocation. 1. Distribution of limited forces and resources for employment among competing requirements. 2. The temporary transfer of forces to meet the operational demand of combatant commanders, including rotational requirements and requests for capabilities or forces (unit or individual) in response to crisis or emergent contingencies. (Approved for incorporation into the DOD Dictionary.)

apportionment. The quantities of force capabilities and resources provided for planning purposes only, but not necessarily an identification of the actual forces that may be allocated for use when a plan transitions to execution. (Approved for incorporation into the DOD Dictionary.)

assumption. A specific supposition of the operational environment that is assumed to be true, in the absence of positive proof, essential for the continuation of planning. (Approved for incorporation into the DOD Dictionary.)

augmentation forces. None. (Approved for removal from the DOD Dictionary.)

available-to-load date. None. (Approved for removal from the DOD Dictionary.)

base plan. A type of operation plan that describes the concept of operations, major forces, sustainment concept, and anticipated timelines for completing the mission without annexes or time-phased force and deployment data. Also called **BPLAN.** (DOD Dictionary. SOURCE: JP 5-0)

branch. 1. A subdivision of any organization. 2. A geographically separate unit of an activity, which performs all or part of the primary functions of the parent activity on a smaller scale. 3. An arm or service of the Army. 4. The contingency options built into the base plan used for changing the mission, orientation, or direction of movement of a force to aid success of the operation based on anticipated events, opportunities, or disruptions caused by enemy actions and reactions. (DOD Dictionary. SOURCE: JP 5-0)

campaign. A series of related operations aimed at achieving strategic and operational objectives within a given time and space. (Approved for incorporation into the DOD Dictionary.)

campaign plan. A joint operation plan for a series of related major operations aimed at achieving strategic or operational objectives within a given time and space. (DOD Dictionary. SOURCE: JP 5-0)

campaign planning. None. (Approved for removal from the DOD Dictionary.)

C-day. The unnamed day on which a deployment operation commences or is to commence. (DOD Dictionary. SOURCE: JP 5-0)

center of gravity. The source of power that provides moral or physical strength, freedom of action, or will to act. Also called **COG.** (DOD Dictionary. SOURCE: JP 5-0)

coalition. None. (Approved for removal from the DOD Dictionary.)

combat support agency. A Department of Defense agency so designated by Congress or the Secretary of Defense that supports military combat operations. Also called **CSA.** (DOD Dictionary. SOURCE: JP 5-0)

commander's estimate. The commander's initial assessment in which options are provided in a concise statement that defines who, what, when, where, why, and how the course of action will be implemented. (Approved for incorporation into the DOD Dictionary.)

completeness. The plan review criterion for assessing whether operation plans incorporate major operations and tasks to be accomplished and to what degree they include forces required, deployment concept, employment concept, sustainment concept, time estimates for achieving objectives, description of the end state, mission success criteria, and mission termination criteria. (Approved for incorporation into the DOD Dictionary.)

concept of operations. A verbal or graphic statement that clearly and concisely expresses what the commander intends to accomplish and how it will be done using available resources. Also called **CONOPS.** (Approved for incorporation into the DOD Dictionary.)

concept plan. An operation plan in an abbreviated format that may require considerable expansion or alteration to convert it into a complete operation plan or operation order. Also called **CONPLAN.** (Approved for incorporation into the DOD Dictionary.)

constraint. In the context of planning, a requirement placed on the command by a higher command that dictates an action, thus restricting freedom of action. (Approved for incorporation into the DOD Dictionary.)

contingency. A situation requiring military operations in response to natural disasters, terrorists, subversives, or as otherwise directed by appropriate authority to protect United States interests. (Approved for incorporation into the DOD Dictionary.)

contingency plan. A branch of a campaign plan that is planned based on hypothetical situations for designated threats, catastrophic events, and contingent missions outside of crisis conditions. (Approved for incorporation into the DOD Dictionary.)

course of action. 1. Any sequence of activities that an individual or unit may follow. 2. A scheme developed to accomplish a mission. Also called **COA.** (Approved for incorporation into the DOD Dictionary.)

crisis action planning. None. (Approved for removal from the DOD Dictionary.)

critical capability. A means that is considered a crucial enabler for a center of gravity to function as such and is essential to the accomplishment of the specified or assumed objective(s). (DOD Dictionary. SOURCE: JP 5-0)

critical requirement. An essential condition, resource, and means for a critical capability to be fully operational. (DOD Dictionary. SOURCE: JP 5-0)

critical vulnerability. An aspect of a critical requirement which is deficient or vulnerable to direct or indirect attack that will create decisive or significant effects. (DOD Dictionary. SOURCE: JP 5-0)

culminating point. The point at which a force no longer has the capability to continue its form of operations, offense or defense. (DOD Dictionary. SOURCE: JP 5-0)

current force. The actual force structure and/or manning available to meet present contingencies. (DOD Dictionary. SOURCE: JP 5-0)

date-time group. None. (Approved for removal from the DOD Dictionary.)

decision. In an estimate of the situation, a clear and concise statement of the line of action intended to be followed by the commander as the one most favorable to the successful accomplishment of the assigned mission. (DOD Dictionary. SOURCE: JP 5-0)

decision point. A point in space and time when the commander or staff anticipates making a key decision concerning a specific course of action. (DOD Dictionary. SOURCE: JP 5-0)

decisive point. A geographic place, specific key event, critical factor, or function that, when acted upon, allows commanders to gain a marked advantage over an enemy or contribute materially to achieving success. (Approved for incorporation into the DOD Dictionary.)

deliberate planning. None. (Approved for removal from the DOD Dictionary.)

deployment order. 1. A directive for the deployments of forces for operations or exercises. 2. A directive from the Secretary of Defense, issued by the Chairman of the Joint Chiefs of Staff, that authorizes the transfer of forces between combatant commanders, Services, and Department of Defense agencies and specifies the authorities the gaining combatant commander will exercise over the specific forces to be transferred. Also called **DEPORD.** (Approved for incorporation into the DOD Dictionary.)

deployment planning. Operational planning directed toward the movement of forces and sustainment resources from their original locations to a specific operational area for conducting the operations contemplated in a given plan. (Approved for incorporation into the DOD Dictionary.)

deterrent options. None. (Approved for removal from the DOD Dictionary.)

dispersion. 1. The spreading or separating of troops, materiel, establishments, or activities, which are usually concentrated in limited areas to reduce vulnerability. (JP 5-0) 2. In chemical and biological operations, the dissemination of agents in liquid or aerosol form. (JP 3-41) 3. In airdrop operations, the scatter of personnel and/or cargo on the drop zone. (JP 3-17) 4. In naval control of shipping, the reberthing of a ship in the periphery of the port area or in the vicinity of the port for its own protection in order to minimize the risk of damage from attack. (DOD Dictionary. SOURCE: JP 4-01.2)

employment. The strategic, operational, or tactical use of forces. (DOD Dictionary. SOURCE: JP 5-0)

essential task. A specified or implied task an organization must perform to accomplish the mission. (Approved for incorporation into the DOD Dictionary.)

estimate. 1. An analysis of a foreign situation, development, or trend that identifies its major elements, interprets the significance, and appraises the future possibilities and the prospective results of the various actions that might be taken. 2. An appraisal of the capabilities, vulnerabilities, and potential courses of action of a foreign nation or combination of nations in consequence of a specific national plan, policy, decision, or contemplated course of action. 3. An analysis of an actual or contemplated clandestine operation in relation to the situation in which it is or would be conducted to identify and appraise such factors as available as well as needed assets and potential obstacles, accomplishments, and consequences. (Approved for incorporation into the DOD Dictionary.)

execute order. 1. An order issued by the Chairman of the Joint Chiefs of Staff, at the direction of the Secretary of Defense, to implement a decision by the President to initiate military operations. 2. An order to initiate military operations as directed. Also called **EXORD.** (DOD Dictionary. SOURCE: JP 5-0)

execution planning. None. (Approved for removal from the DOD Dictionary.)

feasibility. The plan review criterion for assessing whether the assigned mission can be accomplished using available resources within the time contemplated by the plan. (Approved for incorporation into the DOD Dictionary.)

flexible deterrent option. A planning construct intended to facilitate early decision making by developing a wide range of interrelated responses that begin with deterrent-oriented actions carefully tailored to create a desired effect. Also called **FDO.** (DOD Dictionary. SOURCE: JP 5-0)

flexible response. The capability of military forces for effective reaction to any enemy threat or attack with actions appropriate and adaptable to the circumstances existing. (DOD Dictionary. SOURCE: JP 5-0)

force planning. 1. Planning associated with the creation and maintenance of military capabilities by the Military Departments, Services, and United States Special Operations Command. 2. In the context of joint planning, it is an element of plan development where the supported combatant command, in coordination with its supporting and subordinate commands determines force requirements to accomplish an assigned mission. (Approved for incorporation into the DOD Dictionary.)

force sourcing. The identification of the actual units, their origins, ports of embarkation, and movement characteristics to satisfy the time-phased force requirements of a supported commander. (DOD Dictionary. SOURCE: JP 5-0)

fragmentary order. An abbreviated operation order issued as needed to change or modify an order or to execute a branch or sequel. Also called **FRAGORD.** (Approved for incorporation into the DOD Dictionary.)

global campaign plan. Primary means by which the Chairman of the Joint Chiefs of Staff or designated combatant commander arranges for unity of effort and purpose and through which they guide the planning, integration, and coordination of joint operations across combatant command areas of responsibility and functional responsibilities. Also called **GCP.** (Approved for inclusion in the DOD Dictionary.)

governing factors. None. (Approved for removal from the DOD Dictionary.)

H-hour. 1. The specific hour on D-day at which a particular operation commences. (JP 5-0) 2. In amphibious operations, the time the first landing craft or amphibious vehicle of the waterborne wave lands or is scheduled to land on the beach, and in some cases, the commencement of countermine breaching operations. (DOD Dictionary. SOURCE: JP 3-02)

implementation. Procedures governing the mobilization of the force and the deployment, employment, and sustainment of military operations in response to execution orders issued by the Secretary of Defense. (DOD Dictionary. SOURCE: JP 5-0)

implied task. In the context of planning, a task derived during mission analysis that an organization must perform or prepare to perform to accomplish a specified task or the mission, but which is not stated in the higher headquarters order. (Approved for incorporation into the DOD Dictionary.)

indicator. 1. In intelligence usage, an item of information which reflects the intention or capability of an adversary to adopt or reject a course of action. (JP 2-0) 2. In operations security usage, data derived from friendly detectable actions and open-source information that an adversary can interpret and piece together to reach conclusions or estimates of friendly intentions, capabilities, or activities. (JP 3-13.3) 3. In the context of assessment, a specific piece of information that infers the condition, state, or existence of something, and provides a reliable means to ascertain performance or effectiveness. (JP 5-0) (Approved for incorporation into the DOD Dictionary.)

Joint Operation Planning and Execution System. None. (Approved for removal from the DOD Dictionary.)

joint planning. Planning activities associated with military operations by combatant commanders and their subordinate commanders. (Approved for replacement of "joint operation planning" and its definition in the DOD Dictionary.)

joint planning and execution community. Those headquarters, commands, and agencies involved in the training, preparation, mobilization, deployment, employment, support, sustainment, redeployment, and demobilization of military forces assigned or committed to a joint operation. Also called **JPEC.** (DOD Dictionary. SOURCE: JP 5-0)

joint planning group. A planning organization consisting of designated representatives of the joint force headquarters principal and special staff sections, joint force components (Service and/or functional), and other supporting organizations or agencies as deemed necessary by the joint force commander. Also called **JPG.** (DOD Dictionary. SOURCE: JP 5-0)

joint planning process. An orderly, analytical process that consists of a logical set of steps to analyze a mission, select the best course of action, and produce a campaign or joint operation plan or order. Also called **JPP.** (Approved for the replacement of "joint operation planning process" and its definition in the DOD Dictionary.)

Joint Strategic Capabilities Plan. None. (Approved for removal from the DOD Dictionary.)

Joint Strategic Planning System. One of the primary means by which the Chairman of the Joint Chiefs of Staff, in consultation with the other members of the Joint Chiefs of Staff and the combatant commanders, carries out the statutory responsibilities to assist the President and Secretary of Defense in providing strategic direction to the Armed Forces. Also called **JSPS.** (DOD Dictionary. SOURCE: JP 5-0)

leverage. In the context of planning, a relative advantage in combat power and/or other circumstances against the enemy or adversary across any variable within or impacting the operational environment sufficient to exploit that advantage. (Approved for incorporation into the DOD Dictionary.)

L-hour. 1. The specific hour on C-day at which a deployment operation commences or is to commence. (JP 5-0) 2. In amphibious operations, the time at which the first helicopter or tiltrotor aircraft of the airborne ship-to-shore movement wave touches down or is scheduled to touch down in the landing zone. (DOD Dictionary. SOURCE: JP 3-02)

limitation. An action required or prohibited by higher authority, such as a constraint or a restraint, and other restrictions that limit the commander's freedom of action, such as diplomatic agreements, rules of engagement, political and economic conditions in affected countries, and host nation issues. (Approved for replacement of "operational limitation" in the DOD Dictionary.)

limiting factor. A factor or condition that, either temporarily or permanently, impedes mission accomplishment. (DOD Dictionary. SOURCE: JP 5-0)

line of effort. In the context of planning, using the purpose (cause and effect) to focus efforts toward establishing operational and strategic conditions by linking multiple tasks and missions. Also called **LOE.** (Approved for incorporation into the DOD Dictionary.)

line of operation. A line that defines the interior or exterior orientation of the force in relation to the enemy or that connects actions on nodes and/or decisive points related in time and space to an objective(s). Also called **LOO.** (DOD Dictionary. SOURCE: JP 5-0)

major force. A military organization comprised of major combat elements and associated combat support, combat service support, and sustainment increments. (DOD Dictionary. SOURCE: JP 5-0)

measure of effectiveness. An indicator used to measure a current system state, with change indicated by comparing multiple observations over time. Also called **MOE.** (Approved for incorporation into the DOD Dictionary.)

measure of performance. An indicator used to measure a friendly action that is tied to measuring task accomplishment. Also called **MOP.** (Approved for incorporation into the DOD Dictionary.)

mission statement. A short sentence or paragraph that describes the organization's essential task(s), purpose, and action containing the elements of who, what, when, where, and why. (DOD Dictionary. SOURCE: JP 5-0)

multinational. Between two or more forces or agencies of two or more nations or coalition partners. (DOD Dictionary. SOURCE: JP 5-0)

objective. 1. The clearly defined, decisive, and attainable goal toward which an operation is directed. 2. The specific goal of the action taken which is essential to the commander's plan. (Approved for incorporation into the DOD Dictionary.)

operational approach. A broad description of the mission, operational concepts, tasks, and actions required to accomplish the mission. (Approved for incorporation into the DOD Dictionary.)

operational characteristics. Those military characteristics that pertain primarily to the functions to be performed by equipment, either alone or in conjunction with other equipment; e.g., for electronic equipment, operational characteristics include such items as frequency coverage, channeling, type of modulation, and character of emission. (DOD Dictionary. SOURCE: JP 5-0)

operational design. The conception and construction of the framework that underpins a campaign or operation plan or order. (Approved for incorporation into the DOD Dictionary.)

operational design element. None. (Approved for removal from the DOD Dictionary.)

operational pause. A temporary halt in operations. (DOD Dictionary. SOURCE: JP 5-0)

operational reserve. None. (Approved for removal from the DOD Dictionary.)

operation assessment. 1. A continuous process that measures the overall effectiveness of employing capabilities during military operations in achieving stated objectives. 2. Determination of the progress toward accomplishing a task, creating a condition, or achieving an objective. (Approved for inclusion in the DOD Dictionary.)

operation order. A directive issued by a commander to subordinate commanders for the purpose of effecting the coordinated execution of an operation. Also called **OPORD.** (DOD Dictionary. SOURCE: JP 5-0)

operation plan. A complete and detailed plan containing a full description of the concept of operations, all annexes applicable to the plan, and a time-phased force and deployment list. Also called **OPLAN.** (Approved for incorporation into the DOD Dictionary.)

personnel increment number. None. (Approved for removal from the DOD Dictionary.)

phase. In planning, a definitive stage of a campaign or operation during which a large portion of the forces and capabilities are involved in similar or mutually supporting activities for a common purpose. (Approved for incorporation into the DOD Dictionary.)

plan identification number. None. (Approved for removal from the DOD Dictionary.)

planning factor. A multiplier used in planning to estimate the amount and type of effort involved in a contemplated operation. (DOD Dictionary. SOURCE: JP 5-0)

planning order. A planning directive that provides essential planning guidance and directs the development, adaptation, or refinement of a plan/order. Also called **PLANORD.** (Approved for incorporation into the DOD Dictionary.)

preferred forces. Specific units that are identified to provide assumptions essential for continued planning and assessing the feasibility of a plan. (Approved for inclusion in the DOD Dictionary.)

prepare to deploy order. An order issued directing an increase in a unit's deployability posture and specifying a timeframe the unit must be ready by to begin deployment upon receipt of a deployment order. Also called **PTDO.** (Approved for incorporation into the DOD Dictionary.)

ready-to-load date. None. (Approved for removal from the DOD Dictionary.)

required delivery date. None. (Approved for removal from the DOD Dictionary.)

restraint. In the context of planning, a requirement placed on the command by a higher command that prohibits an action, thus restricting freedom of action. (Approved for incorporation into the DOD Dictionary.)

risk. None. (Approved for removal from the DOD Dictionary.)

scheme of maneuver. The central expression of the commander's concept for operations that governs the development of supporting plans or annexes of how arrayed forces will accomplish the mission. (Approved for incorporation into the DOD Dictionary.)

sequel. The subsequent operation or phase based on the possible outcomes of the current operation or phase. (Approved for incorporation into the DOD Dictionary.)

shortfall. The lack of forces, equipment, personnel, materiel, or capability, reflected as the difference between the resources identified as a plan requirement and those quantities identified as apportioned for planning that would adversely affect the command's ability to accomplish its mission. (Approved for incorporation into the DOD Dictionary.)

specified task. In the context of planning, a task that is specifically assigned to an organization by its higher headquarters. (Approved for incorporation into the DOD Dictionary.)

staff estimate. A continual evaluation of how factors in a staff section's functional area support and impact the planning and execution of the mission. (Approved for inclusion in the DOD Dictionary.)

strategic communication. None. (Approved for removal from the DOD Dictionary.)

strategic concept. None. (Approved for removal from the DOD Dictionary.)

strategic direction. The strategy and intent of the President, Secretary of Defense, and Chairman of the Joint Chiefs of Staff in pursuit of national interests. (Approved for incorporation into the DOD Dictionary.)

strategic estimate. The broad range of strategic factors that influence the commander's understanding of the operational environment and the determination of missions, objectives, and courses of action. (Approved for incorporation into the DOD Dictionary.)

strategic guidance. The written products by which the President, Secretary of Defense, and Chairman of the Joint Chiefs of Staff provide strategic direction. (Approved for inclusion in the DOD Dictionary.)

strategic plan. None. (Approved for removal from the DOD Dictionary.)

subordinate campaign plan. A combatant command prepared plan that satisfies the requirements under a Department of Defense campaign plan, which, depending upon the circumstances, transitions to a supported or supporting plan in execution. (DOD Dictionary. SOURCE: JP 5-0)

supporting plan. An operation plan prepared by a supporting commander, a subordinate commander, or an agency to satisfy the requests or requirements of the supported commander's plan. (DOD Dictionary. SOURCE: JP 5-0)

time-phased force and deployment data. The time-phased force, non-unit cargo, and personnel data combined with movement data for the operation plan, operation order, or ongoing rotation of forces. Also called **TPFDD.** (Approved for incorporation into the DOD Dictionary.)

times. The Chairman of the Joint Chiefs of Staff coordinates the proposed dates and times with the commanders of the appropriate unified and specified commands, as well as any recommended changes when specified operations are to occur (C-, D-, M-days end at 2400 hours Universal Time [Zulu time] and are assumed to be 24 hours long for planning). (DOD Dictionary. SOURCE: JP 5-0)

transportation feasible. A determination made by the supported commander that a draft operation plan can be supported with the identified or assumed transportation assets. (Approved for incorporation into the DOD Dictionary.)

Universal Time. A measure of time that conforms, within a close approximation, to the mean diurnal rotation of the Earth and serves as the basis of civil timekeeping. Also called **ZULU time.** (Approved for incorporation into the DOD Dictionary.)

validate. Execution procedure used by combatant command components, supporting combatant commanders, and providing organizations to confirm to the supported commander and United States Transportation Command that all the information records in a time-phased force and deployment data not only are error-free for automation purposes, but also accurately reflect the current status, attributes, and availability of units and requirements. (DOD Dictionary. SOURCE: JP 5-0)

warning order. 1. A preliminary notice of an order or action that is to follow. 2. A planning directive that initiates the development and evaluation of military courses of action by a commander. Also called **WARNORD.** (Approved for incorporation into the DOD Dictionary.)

Intentionally Blank

JOINT DOCTRINE PUBLICATIONS HIERARCHY

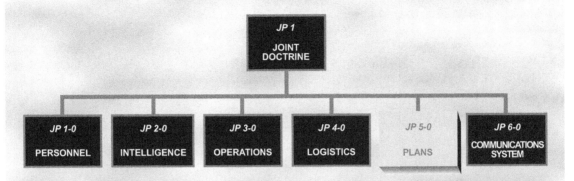

All joint publications are organized into a comprehensive hierarchy as shown in the chart above. **Joint Publication (JP) 5-0** is in the **Plans** series of joint doctrine publications. The diagram below illustrates an overview of the development process:

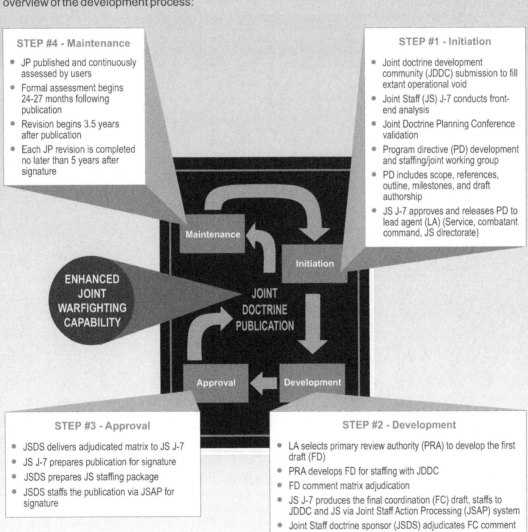

STEP #4 - Maintenance

- JP published and continuously assessed by users
- Formal assessment begins 24-27 months following publication
- Revision begins 3.5 years after publication
- Each JP revision is completed no later than 5 years after signature

STEP #1 - Initiation

- Joint doctrine development community (JDDC) submission to fill extant operational void
- Joint Staff (JS) J-7 conducts front-end analysis
- Joint Doctrine Planning Conference validation
- Program directive (PD) development and staffing/joint working group
- PD includes scope, references, outline, milestones, and draft authorship
- JS J-7 approves and releases PD to lead agent (LA) (Service, combatant command, JS directorate)

STEP #3 - Approval

- JSDS delivers adjudicated matrix to JS J-7
- JS J-7 prepares publication for signature
- JSDS prepares JS staffing package
- JSDS staffs the publication via JSAP for signature

STEP #2 - Development

- LA selects primary review authority (PRA) to develop the first draft (FD)
- PRA develops FD for staffing with JDDC
- FD comment matrix adjudication
- JS J-7 produces the final coordination (FC) draft, staffs to JDDC and JS via Joint Staff Action Processing (JSAP) system
- Joint Staff doctrine sponsor (JSDS) adjudicates FC comment matrix
- FC joint working group

Made in the USA
Coppell, TX
10 February 2022